Springer Polar Sciences

This Series is now being indexed by SCOPUS *Springer Polar Sciences* is an interdisciplinary book series that is dedicated to research in the Arctic, sub-Arctic regions, and the Antarctic. In recent years, the polar regions have received increased scientific and public interest. Both the Arctic and Antarctic have been recognized as key regions in the regulation of the global climate, and polar ecosystems have been identified to be particularly susceptible to the ongoing environmental changes. These changes are having widespread implications for human communities, businesses, and governance systems and are interacting with demographic shifts, globalisation, resource development, cultural change, territorial disputes, and growing calls for self-determination in some regions. Consequently, the international efforts in polar research have been enhanced considerably, and a wealth of new findings is being produced at a growing rate by the international community of polar researchers and those who live in the region.

Springer Polar Sciences aims to present a broad platform that will include state-of-the-art research, bringing together both science, humanities, and perspectives rooted in indigenous and local knowledge to facilitate an exchange of knowledge between the various polar science communities. The series offers an outlet to publish contributions, monographs, edited works, conference proceedings, etc. Topics and perspectives will be broad and will include, but not be limited to, climate change impacts, climate change policy, environmental change, polar ecology, governance, health, economics, indigenous populations, tourism, resource extraction activities, and research design in polar regions. Books published in the series will appeal to scientists, students, polar researchers, community leaders, and policy makers.

Matthew Bampton

The North Atlantic Polar Triangle

Documenting The End of an Epoch

Thanks for the help Lee
Matthew

Springer

Matthew Bampton
Department of Geography-Anthropology
University of Southern Maine
Portland, ME, USA

ISSN 2510-0475 ISSN 2510-0483 (electronic)
Springer Polar Sciences
ISBN 978-3-031-27263-9 ISBN 978-3-031-27264-6 (eBook)
https://doi.org/10.1007/978-3-031-27264-6

This Springer imprint is published by the registered company Springer Nature Switzerland AG
The registered company address is: Gewerbestrasse 11, 6330 Cham, Switzerland

Preface

This book is my attempt to unpack the question of how humans interact with the environment: how we change it and how it changes us. I approach this from the perspective of my training as a geographer, and with 30 years of work teaching and researching in the field behind me. Broadly stated, geographers study the ways in which landscapes are made and how they function. The essential geographical question is: "why is the world arranged in this way"? Any reasonable answer to this question must include a consideration of how humans have influenced processes on the surface of the Earth. This has fascinated me since I was an undergraduate. I study how humans change the world around them and how that interaction has changed through time.

Three laws guide my thinking:

- Commoner's First Law of Ecology: Everything is connected to everything else (Commoner 1971).
- Tobler's First Law of Geography: Everything is related to everything else, but near things are more related than distant things (Tobler 1969).
- Forman and Godron's First Law of Landscape Ecology: An action here and now produces an effect there and then (Forman and Godron 1986).

Not all processes are affected by humans. Plate tectonics, for example, are completely independent of social, cultural, or economic forces. However, a great many other Earth systems are heavily influenced by what people do individually and collectively. Foremost among these at present is human-driven climate change. But to focus only on this would be a mistake. Almost every other Earth surface system is also influenced by humans. A complex web of causation and interdependency links all of these systems to one another. There is now a half-hearted international effort to ameliorate the worst consequences of the climate crisis by managing its causes. The most immediately identifiable of these is the colossal amount of fossil fuel used daily in the developed world, a quantity that has been steadily increasing year on year since the beginning of the Industrial Revolution, in the eighteenth century. A host of ancillary causes include, among other things, changes in global carbon budget, in surface reflectivity, and in evapotranspiration. The immediacy and visibility of the climate crisis has inspired a number of powerful activist movements, including a spectacular youth effort. It has also provided the impetus for a cynical campaign of misinformation, misdirection, and malpractice heavily funded by a few wealthy industrialists and promulgated by paid commentators (Mann 2013). The prominence of climate change in recent years has tended to distract us from other questions of human transformations of the natural environment. There are matching dynamics of transformative human actions in almost every sphere of the natural world. For example, rapidly accelerating extinction rates, deforestation, soil loss, ocean acidification, and the increasingly frequent collapse of complex ecosystems are all aspects of the same general trend: humans are now a forcing agent in a wide array of Earth systems. This tendency is so marked that there has been a credible scientific effort to call time on the current geological epoch, the Holocene, and proclaim the beginning of a new one. The term “Anthropocene” is sometimes used to describe the boundary between the pre-humanized world and the humanized world.

This is a truly extraordinary suggestion. The notion that behavior of a significant number of the Earth’s natural systems is now driven by the behavior of a single species is without historical precedent. The nearest any previous species has come to changing the behavior of the entire planet was during the Archaean, starting around 3.6 billion years ago, when blue-green algae successfully oxygenated the atmosphere. It took them about a billion years. Anatomically modern humans have only been around for something between 350,000 and 195,000 years. This is a tiny fragment of time – one analogy sometimes used to explain this is to imagine Earth history as a journey from one side of the USA to the other. On such a journey, of around 4560 km or 2800 miles, humans only show up in the last 400 m. Yet within this time we humans have managed to completely transform the functioning of several of the Earth’s more vulnerable systems. We have also done this with a fairly sophisticated understanding of what we are doing. Particularly with the last couple of centuries we have developed a good enough idea of how the planet works to measure the impact we are having on it and to predict with some accuracy how our actions will play out over the longer term.

My intention here is to explore this idea in detail by looking at its implications for a single coherent region on the Earth’s surface, the North Atlantic and its surrounding landmasses. This is in part because although our collective understanding of what we are doing is pretty good, individually we don’t have such a great handle on what’s going on. Science advances rapidly, so new knowledge is constantly being added to the encyclopedia. Existing specialist fields are complex and require specialized skill sets to understand. New fields are emerging. Because of the technical complexity of diverse areas of study, and because of the intellectual tricks that sometime accompany them, it can be difficult for outsiders to understand specialized knowledge. I am going to take a single part of the Earth’s surface with which many people have some familiarity. Within this region a number of key environmental processes behave in a closely coupled way. These links impact the behavior of living systems, and it also impacts the ways in which humans interact. I argue that the North Atlantic and its surrounding landmasses are illustrative of changes happening throughout the Earth system as a whole.

I make this argument as the study region I have defined; the North Atlantic Polar Triangle (hereafter NAPT) covers a region that extends from the equator to the pole. Within this region there are major circulation patterns of ocean and atmosphere that both serve to move thermal energy from the equator pole-wards. This combination of circulation is one of the factors that defines the climate of the whole region and therefore also its ecosystems. The basic physical building blocks of this region have been moving into place for around 65 million years. Changes in the way it functions are frequently derived either from things that are happening within the system as it stands or from completely outside the system. For almost all of the past 195,000 years, the geography of the region proved an insurmountable barrier to the westward migration of humans, who only devised strategies to overcome this challenge around 1000 years ago. Consequently, it was also the place where the human circumambulation of the world was finally completed. With the first encounter between a Norse settler and a Greenlandic hunter, humans had finally completed the journey around all 360° of the world's longitude. Greenland was the place where the eastward and westward migrations of humans finally closed the circle. This bridging of the Atlantic marked a connection between what had hitherto been two parallel histories. Although the cultures on each side of the Atlantic had developed independently since humans crossed the Baring Straits, there are some striking similarities between their trajectories, particularly in the post-ice age period: in both cases there were significant cultural changes, population expansion, and the development of domestication, agriculture, ceramics, and metallurgy.

To put recent human forcing on natural systems into some sort of historical context, it is useful to understand something about chronostratigraphy, or the study of geological time. Chronostratigraphers, under the rubric of the International Commission on Stratigraphy (ICS), divide the 4.6 billion years of Earth history into a nesting hierarchy of divisions and subdivisions. These range from Eons, through Periods, Epochs, and Ages. The divisions reflect changes in the prevailing conditions on the Earth's surface. Over time the general characteristics of atmosphere, climate, hydrosphere, biosphere, and tectonics create distinctive signatures in the geological record. Rocks formed in the distant past, when the Earth was a high-temperature volcanically active planet with very little surface liquid water, and where life was restricted to cyanotic algae, are very different from rocks formed in shallow tropical oxygen-rich oceans with a teeming, complex, and diverse ecosystem.

It is important to remember that these divisions of time are no more or less objectively valid than the divisions historians create to understand the complexities of human history, or that archaeologists create to understand prehistory. They are simply a useful way to split large and complex swathes of time into manageable chunks. The divisions tend to center on general tendencies, but they usually have rather blurry boundaries, and they are rarely universally applicable. So, for example, the "agricultural revolution" happened in several places, at several different times, and in several different ways starting around 8000 years ago. There seem to have been a few false starts; in some places people experimented with domestication, and then for one reason or another abandoned the idea. Our knowledge of this process is constantly changing, as we learn more about how and where agriculture was invented or established. Neither the complexity of its pattern in time or space diminishes its significance for human evolution. There was a time when everyone on Earth obtained all their food from wild sources. In the present-day world, only a vanishing and tiny proportion of humans still wild harvest all or even most of their nutrition. Nearly everybody else, and certainly everyone reading these words, gets nearly every calorie they consume from some domesticated plant or animal. The boundary between non-agricultural and agricultural humanity is complex and extends across a lot of time and space. But like chronostratigraphic boundaries it can be defined, and it is identifiable, significant, and consequential.

Chronostratigraphic boundaries are defined by a datable and globally recognizable signature in rock, sediment, or ice that signifies the historical onset of a distinctive set of physical, chemical, or biological conditions. There are a number of factors that can create such signatures. For example, tectonic arrangement – the position of the Earth's continents at any given point in history – can profoundly affect many processes. We tend to be aware of continental movement as a slow process that only manifests itself meaningfully within a human lifespan in dramatic events such as earthquakes, volcanoes, and tsunamis. In chronostratigraphic terms, the important tectonic questions are much bigger: how many continents are there? One big one, or several smaller ones? Where are they located? In the tropics, or at the poles? How are the oceans arranged? What is their orientation? As with the example of human historical changes, things can get quite complex. If all of the continental crust is sitting over the South Pole, and the rest of the Earth's surface is water, many of the rules about how surface processes work will be altered in some pretty important ways. Atmospheric circulation functions differently absent the friction of landmasses and the differential heating of land and water. Ocean circulation is directed differently as blocking landmasses are differently distributed. Evolutionary possibilities unfold differently as interactions between organisms and their environment are differently constrained.

The signature of climatic and atmospheric influences on surface processes of sufficient magnitude to define chronostratigraphic boundaries can sometimes be seen in rocks, sediments, ice core data, and speleothems. By far the most reliable indicator used to identify such boundaries is, however, ecology. Prior to the development of a broadly accepted history of the earth, or comprehensive theory of evolution, the differences between fossil species visible to the naked eye provided the framework for a theory of geochronology. Because living things are so responsive to their surroundings, it is the changes in the fossil record that most typically indicate extensive changes in other sets of processes, and that therefore most frequently serve to divide geological time up into convenient categories and sub-categories. The four biggest divisions of geological history, the Eons, reflect this. These are the Hadean (hell-like), the Archaean (ancient), the Proterozoic (beginnings of life), and the Phanerozoic (visible life). There is now pretty good evidence to indicate that life was already well-established by the Archaean and was rolling along nicely through the Proterozoic. The beginning of the Phanerozoic witnessed the so-called "Cambrian Explosion" when complex multicellular life blossomed into an ever-increasing and ever more complex array of kingdoms, phyla, classes, orders, families, genera, and species. The half-billion years that has so far comprised the balance of this most recent Eon have seen an acceleration, diversification, and proliferation of life-strategies that sharply contrasts with the four preceding "boring billions."

The notion that humans are now a causal factor in pattern formation at this scale is extraordinary. Positioning people as a forcing agent anywhere within the nesting scales of geological time smacks of sublime hubris. Named chronostratigraphic times stages at any scale represent huge swaths of time. The boundaries that separate them are spectacular catastrophes often unfolding over millennia. When such changes are considered in comparison to the scale of human experience, it is hard to imagine that we are players in this league. But as Commoner, Tobler, Forman, and Godron respectively observe, everything is related, proximity matters, and effects propagate through time and space. There is now a compelling body of evidence to indicate that humans are now manipulating a diverse array of Earth systems to such an extent that we are now passing through a chronostratigraphic boundary. The magnitude of the change, and consequently the rank of the next time remains in question. Still, we are undoubtedly witnessing changes of an unprecedented order within recorded history, and possibly within all of human history.

References

Commoner B (1971) The closing circle: nature, man, and technology. Alfred A. Knopf, New York

Forman RTT, Godron M (1986) Landscape ecology. Wiley, New York

Mann ME (2013) The hockey stick and the climate wars: dispatches from the front lines. Columbia University Press, New York

Tobler W (1970) A computer movie simulating urban growth in the Detroit region. Econ Geogr 46(Supplement):234–240

Acknowledgements

A number of friends and colleagues helped me conceive, undertake, and complete this book. My ongoing conversations with fellow Earth scientists Mark Swanson and Lee Slater were particularly important. Others whose input moved the work along in various ways include Alice Kelley, Joe Kelley, Joe Szakas, and Judy Tupper. A Fulbright Fellowship at the University of Edinburgh supported my initial research. A USM sabbatical provided time for a significant part of the final writing effort. Rachel Hale, my research assistant, provided invaluable support. Erin Little, Mary Minehan, and Stephen Purvis, each in their various ways, encouraged me and helped me develop my thoughts. My daughters, Ariel, Freya, and Heloise, continue to inspire my work. Most importantly, none of this would have been possible without the constructive criticism, support, and encouragement of my wife Lisa Page. Any errors are, of course, entirely mine.

Contents

Chapter 1
Magnitude, Frequency, and Change in Earth Systems

Purpose of the Book, Structure, and Foundational Concepts

Abstract Five foundational concepts support the arguments made in this book: (1) the conventions of geological time and chronostratigraphy, (2) background extinction rates, major extinction events, and possible kill-mechanisms, (3) climate change (4) the geography of the study area, and (5) the recently proposed new chronostratigraphic unit the Anthropocene.

Keywords Geological time · Chronostratigraphy · Extinction · Anthropocene

1.1 Introduction

Contemporary humans are exerting an influence on natural systems that is profound, far-reaching, and without historical precedent. Our collective impact is now so marked that there is a proposal among some Earth scientists to designate a new division of geological time, the Anthropocene. My intention here is to investigate this idea. I will make the case that humans are now a major forcing agent of planetary process. The changes we have wrought can be positioned within the framework of geological history. The magnitude of this change, and whether we are creating a new epoch, or we are creating a boundary between the Holocene, which follows the Last Glacial Maximum, and some currently unknown future state is one of the central ideas I will explore here.

The geographical focus of this book is a region I have designated as the North Atlantic Polar Triangle (NAPT). This is a convenient area in which to explore the question of how humans have progressively increased their influence on global systems, as it provides a manageable unit within which human and natural systems can be tracked over the entire Holocene. The NAPT (Fig. 1.1), extending from 90°N to 0° covers about a third of the northern hemisphere. Within this space is a diverse suite of physical environments from tropical rainforests, through deserts, savannahs, temperate forests, mountains, tundra, and the Arctic. Within the same space during the Holocene human society has passed from Paleolithic adaptations through agriculture and feudal society, to the Industrial Revolution, and beyond. This has

M. Bampton, *The North Atlantic Polar Triangle*, Springer Polar Sciences,
https://doi.org/10.1007/978-3-031-27264-6_1

Fig. 1.1 The North Atlantic polar triangle

happened at different paces in different places, sometimes independently, and sometimes because of cultural diffusion, population displacement, or colonization. The Atlantic stood as one of the last geographical barriers to human cultural diffusion. Humans had managed to overcome almost all other barriers, even walking the long way round the world, from Africa to Greenland, before finally traversing the Atlantic in 1021. With this voyage Norsemen finally finished the human journey circumscribing all 360° of longitude.

A discussion limited to the geographical scope of the NAPT obviously ignores important physical and cultural dynamics elsewhere in the world. I have chosen to restrict this study in space because I believe a broader geographical scope would be overwhelmingly complex and would be doomed to failure. I have favored a manageable case study over an overwhelming comprehensive review.

Scientifically reliable predictions of anthropogenic changes to foundational natural systems shape every reasonable discussion of the human condition. Evidence of human transformation of Earth surface processes is so incontrovertible and widespread that at the beginning of this century it prompted the proposal to designate a new geological epoch, the Anthropocene, to supersede the Holocene (Crutzen and Stoermer 2000; Waters et al. 2014). The challenges now confronting humans are so serious that even otherwise unfailingly optimistic commentators recognize the looming existential crisis (Pinker 2021). In the first quarter of the twenty-first century, we are faced with the prospect of knowingly, fundamentally, and irrevocably changing the functioning of enough surface systems to precipitate at least a catastrophic human population collapse, and possibly even our own extinction. This is a unique and interesting moment in planetary history, as we are the first species to simultaneously engineer such a remarkable feat, and also to accurately document its progress. This book explores two questions arising from this remarkable situation: how does it fit into the context of geological history, and how does it fit into the context of human history?

My intent is to position the current discussion of the Anthropocene, a notion that now extends far beyond the rarified realm of chronostratigraphy, in a more rigorous consideration of what we know of system change on the Earth's surface. By exploring past dynamics in ecological, and climate systems I will argue that in one perspective late Holocene trends are really not that big a deal: there have been far more dramatic changes in the past. Within the Phanerozoic temperatures have fluctuated widely, and sea levels have at various times been much higher and much lower than they are at present. Past extinction events have been much more dramatic than anything we are currently witnessing, with some events almost completely wiping life from the face of the planet. In the shorter timeframe of anatomically modern humans, or even merely the Holocene, population collapse within and beyond the geographical constraints of the NAPT, has occurred numerous times. Our current situation is at the at the zenith of human knowledge, population, longevity, wealth, and dominion over nature. Following past catastrophes in prehumen times ecosystems have equilibrated and organisms have evolved and adapted. Within human times populations have recovered, lifeways have changed, and new societies have developed. While these processes have so far filled the gaps left by natural and human disaster, the time of change is extremely unpleasant, and for entire species and cultures, often fatal. And there are no guarantees about who or what will survive, and what the next system will look like.

The book is organized into three parts. Part 1 has two chapters. This chapter delineates the study region and outlines some of the essential theoretical underpinnings of Earth system science. Chapter 2 provides an overview of the history of the NAPT prior to the Holocene. Part 2 has three chapters. These cover the three major chronostratigraphic divisions of the Holocene: the Greenlandian, the Northgrippian, and the Meghalayan. Part 3 comprises the final chapter, outlining the concept of the Anthropocene as a chronostratigraphic boundary, rather than a new geological age, and an epilogue posing the question "what now"?

The "deep time" historical perspective does not however leave room for complacency. Past human population collapses have been terrible calamities. Subsequent recoveries do not alleviate the suffering and tragedy of these events. While we may draw comfort from the fact that we are now collectively enjoying a "golden era" of prosperity and technology, all available evidence indicates our species' periodic episodes of disaster are extremely unpleasant. For example, in the Americas, following the Columbian impact, an estimated 98% of the indigenous population perished. It is difficult to celebrate the species-level biological resilience of humanity, now enjoying an unprecedented demographic boom on this recently depopulated territory, without at least considering its cost in lives and cultures lost. Maybe Homo sapiens will survive another civilizational collapse, but it will be an unpleasant process, and there are no guarantees about who and what will be lost or saved.

Still more sobering is the perspective of major systemic change and subsequent survival working in the perspective geological time. Both extinction and evolution are driven by a complex set of interactions between competition, species characteristics, environmental conditions, and dumb luck. However combined, these factors can rapidly and unpredictably clean house on the planetary scale. In such circumstances "some innocents 'scape not the thunderbolt": however well adapted anyone is to today's conditions, tomorrow's challenges can be fatal. In the wake of disaster, evolutionary recovery is typically measured in hundreds of thousands, or millions of years, and involves false starts, blind alleys, and varied fortunes beyond the control of any organism sentient or otherwise.

1.2 Geological Time and Chronostratigraphy

To talk about the history of the Earth it is first useful to understand how geologists divide it up. By current reckoning the Earth has been around for about 4.6 billion years. It can be difficult to grasp the magnitude of so much time even if you are familiar with the geological column. The brief span of a human life makes our personal experience hopelessly inadequate when confronted with the reality of deep time. On the scale of planetary history, the foundational principles of physical, biological, and Earth sciences manifest in unfamiliar ways. The relationships and processes that we observe day-to-day play out in unexpected ways over extremely long stretches of time. Continents move around, collide, and break apart; mountains fold upwards and then are worn away; the axis of the planet wobbles, and the Earth's magnetic polarity reverses; the atmosphere's composition, structure and circulation change, ocean currents shift around, and; an extraordinary and diverse array of living things come and go on the planetary stage.

The resulting time-divisions are an intellectual convenience, and are as much a human construct as are the divisions historians impose on human history. The Earth has only one history. Although this doesn't diminish their validity or usefulness, it does mean that the time-slices and their boundaries are constantly being re-examined and are subject to change in the light of new information. This noted, three general principles hold:

In order to manage geological time geologists, use a nesting hierarchy of named slices: eons, eras, periods, epochs, and ages. The business of organizing and managing this system falls to the science of specialist field of chronostratigraphy. For example, about two thirds of the way through the Phanerozoic Eon, during the Mesozoic Era, the Cretaceous Period, occurs, between 145 million years ago (Mya) and 66 Mya. It is divided into two epochs, and six ages. The smaller divisions of time always fit inside the larger divisions of time, either by subdividing them, or by sharing their definitive boundaries. The Campanian and Maastrichtian Ages make up the last 17.6 million years of the Upper Cretaceous Epoch, and also of the Cretaceous Period. The end of the Maastrictian, the end of the Upper Cretaceous, and the end of the Cretaceous all occur at the same time, and mark the beginning of the Paleogene Period (Fig. 1.2). Each division in time, wherever it sits in the hierarchy, is defined by some clear mark in the geological record. This is often a significant shift in the nature of ecosystems, indicated by widespread extinctions, followed by a burst of evolutionary activity. Such changes are visible in the relative abundance of identifiable fossilized organisms. Other indicators are sudden changes in the rock record itself, or in proxy indicators that show changes in atmospheric or marine temperature and composition.

- Any boundary between two or more time slices indicates changes on the Earth that are qualitative and fundamental.
- During any named time-slice at any level in the hierarchy processes on the Earth operate to change things quantitatively and incrementally.
- The higher in the hierarchy the boundary occurs the bigger the changes that define it.

One of the fundamental organizing principles of chronostratigraphy is the notion of unconformity. This idea was first articulated by the eighteenth-century geologist James Hutton. Geological unconformities, visible breaks in the sequence of events recorded in the rock record, are evidence of the episodic nature of Earth processes.

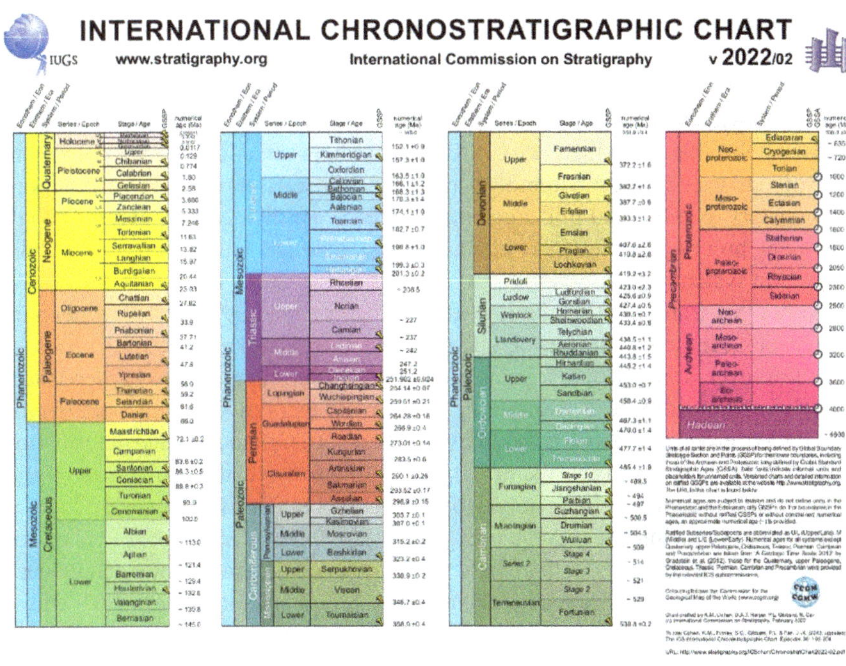

Fig. 1.2 International chronostratigraphic chart (https://stratigraphy.org/chart)

Each rock, regardless of whether it is igneous, sedimentary, or metamorphic, has unique characteristics of physical composition and a unique place in Earth history. The constituent minerals, and the way in which they are assembled are a direct consequence of the prevailing conditions in which the rock was formed and when it happened. In the case of an igneous rock the chemical composition of its minerals, and the size and arrangement of the crystals they form, indicates the type of source magma, its melting point, its depth within the crust, and the time it took to cool and solidify. A sedimentary rock preserves a record of its source material, sometimes another rock, sometimes an organism, the physical conditions of formation, and the type of diagenesis (lithification) by which it was formed. A metamorphic rock preserves a record of the source rock from which it was formed, and the force vectors, temperatures, and pressures that transformed it into new materials. Because different rocks tell the story of different conditions of formation, the juxtaposition of any two different rock types in situ indicates a change in conditions of some kind. For example, a series of different sand layers in a cross-section through a beach dune shows how changes in wind speed have sorted grain sizes over time. When the wind speeds up bigger grains move, and when it slows they fall: a layer of fine sand indicates a period of very slow wind, a layer of coarse sand indicates faster wind.

The previously mentioned Campanian-Maastrichtian Boundary Event, dividing two ages, is defined by global cooling, the tectonic opening of an Equatorial Atlantic Seaway, and a perturbation of the global carbon cycle. By contrast, the end of the Maastrictian, which coincides with the end of the Cretaceous Period, is defined by the K-T extinction, in which three quarters of the plant and animal species on the planet were eliminated following the meteor impact that created what Walter Alvarez memorably named "The Crater of Doom", more formally known as the Chicxulub crater (Alvarez 2008).

This notwithstanding, regardless of the age or sequencing of rock units, the general rule holds that when there are big differences between adjacent rock units there were big changes in the environment in which they were formed. The chapters of Earth history as preserved in the geological record are divided between periods of consistent process driving incremental change that are punctuated by periodic changes in system state.

Within this framework geological features can be read as logical statements in the bedrock. At the simplest level they illustrate a necessary sequence of events. A lava flow over a sedimentary layer must have occurred after it was deposited. Upended sedimentary beds overlain by horizontal layers of sediment were deposited, folded, eroded, and then covered by new deposits. Orientation and deformation of folds indicate strength and vector of force. While some relationships are straightforward, interpretation is frequently a lot more complex. Thrust-faults and overturn-folds can upend timelines. A new exposure can force a reinterpretation of the landscape. There are gaps in the record when geological events obliterate evidence of the environments that preceded them. There are times and places in which the record is sparse or absent: processes do not affect all of the Earth's surface in the same way, to the same extent, or at the same time. In the field the record is frequently ambiguous, inconsistent, and incomplete.

The earliest successful attempts to interpret Earth history in this way, such as William Smith's 1815 geological map of England, relied on the exhaustive cataloging of similarities and differences between rock units, and fossil assemblages. Where consistencies existed a consistent set of conditions in time and space was hypothesized. Big differences in the rock record marked significant changes in process and conditions. One of the challenges presented by this approach was to extrapolate over large areas of space. Local unconformities are not necessarily indicators of global changes. Although the earliest efforts to create general rules of geological sequencing were hampered by limited data, the accumulation of new information, the improvement of techniques, and three centuries of study have all helped to refine our collective understanding of the Earth's history. These methods still comprise much of the work of practicing field geologists. Throughout two guiding principles, first articulated by Hutton, have held. These are summarized by Jackson:

> The first Law said that in a sequence of beds of rock, those that lie on top are younger than those below, unless there is clear evidence to suggest that the whole succession has overturned. Such reversals could be caused by folding or faulting in the rocks. [The] second Law said that each bed each contained a distinctive fossil assemblage (Jackson 2006: 128)

The geological time-scale within which the resulting historical interpretation is managed is now supervised by the International Commission on Stratigraphy. They use five categories for describing rocks:

1. Unconformity-bounded units – bodies of rock bounded above and below by significant discontinuities in the stratigraphic succession.
2. Lithostratigraphic units – units based on the lithologic properties of the rock bodies.
3. Biostratigraphic units – units based on the fossil content of the rock bodies.
4. Magnetostratigraphic polarity units – units based on changes in the orientation of the remnant magnetization of the rock bodies.
5. Chronostratigraphic units – units based on the time of formation of the rock bodies.

The relative dating possible using the law of superposition is now supplemented by objective dating techniques that independently determine the age of rocks in years before the present. Most of these are based on the radioactive decay of one or other of the isotopes in the rock. More recent materials can be objectively dated using other techniques such as optically stimulated luminescence (OSL), obsidian hydration, and prior calcite precipitation. The improved dates have significantly refined the geological time-scale, and enabled a more nuanced interpretations of Earth history.

In practice most of the time the first four categories end up defining the fifth. That is, if a rock has distinctive characteristics of composition (1), if it is very different from other rocks above and below it (2), if it has very different fossils (3), or if it has a radically different magnetic signature (4), then it was probably formed at a different time (5).

The divisions of geological time are in one sense just a convenience – the Earth has only one history. Yet unlike the periods and ages of human history, the geochronological divisions of time are less subjective. Debates about the divisions of human history get tangled in questions of legitimacy such as their universality in time and space. For example, there was neither a Bronze Age nor a Renaissance in the Americas. Even in the twenty-first century defining cultural events are still nearly always regionally constrained: election results, civil wars, and "global financial catastrophes" are all restricted in scope. In the case of geological time the debate is easier to frame. To designate a new chapter at any level in the geochronological hierarchy a set of pretty well-defined criteria must be met. Most importantly, a "Global Boundary Stratotype Section and Point" or GSSP must be identified. The International Commission on Stratigraphy describes what conditions must be met to qualify as a GSSP, and so to define a new chronostratigraphic boundary, at whatever level:

- The lower boundary has to be defined using a primary marker (usually first appearance datum of a fossil species).
- A GSSP has to define the lower boundary of a geologic stage.
  - There should also be secondary markers (other fossils, chemical, geomagnetic reversal).
  - The horizon in which the marker appears should have minerals that can be radiometrically dated.
  - The marker has to have regional and global correlation in outcrops of the same age
  - The marker should be independent of facies.
- The outcrop has to have an adequate thickness
- The outcrop should be unaffected by tectonic and sedimentary movements, and metamorphism
- Sedimentation has to be continuous without any changes in facies
- The outcrop has to be accessible to research and free to access.

In plain language this means that a named time-slice has to be constrained by a lower boundary, which also serves as a definitive terminus for the previous slice. This must be visible in a rock outcrop somewhere. There has to be a clearly identifiable fossil marker in the outcrop which has not previously been used to define a different boundary, and it must be datable. It should be visible, and undisturbed by tectonics. And it has to be accessible to researchers.

Not all boundaries meet all of these stringent requirements, and not all boundaries are undisputed. The International Commission on Stratigraphy is continually revising and refining their classification schema as new discoveries are made. However, in general terms the principles used to separate one named time from another are accepted. When there are worldwide changes in the way natural systems work that change the character of rocks and living systems for long enough to show up in the geological record a new time slice is defined. Smaller systemic changes create lower-ranking divisions, bigger changes are higher in the hierarchy. Systemic consistency through time is the age, systemic collapse and discontinuity is the boundary.

The best documented boundary to date is the most recent, the end of the LGM, and beginning of the Holocene. Pielou provides an account of the return of life to North America following the retreat of the glaciers (Pielou 2008). Her detailed discussion of the mechanisms by which surviving outliers of species migrated into suddenly hospitable landscapes, how populations became either isolated, or newly connected, and how a host of other interactions and developments unfolded, gives a powerful impression of landscape evolution following a major boundary event. We can deduce that every preceding boundary event, large or small, underwent proportionately scaled and comparable changes. In the literature there are numerous accounts of how environments, ecosystems, and individual taxons responded, or failed to respond, to such changes in the more distant past. However, Pielou's book records the behavior of an entire continental system, and presents a series of clearly articulated hypotheses and models. For instance, her analysis of constraints to plant refugia – small areas where individuals survive harsh times – and migration out of these places following habitat change – can be scaled and transferred to numerous other cases, past and present.

There are, of course, gaps in this record. Likewise, there are enigmas, contradictions, and undoubtedly mistakes. Despite these "wrinkles in time" the grand sweep of Earth history is now fairly well mapped out. Whether you define "now" geologically (the current age), in human terms (the current moment), or anywhere in between, the landscape is a single unique instant in the Earth's history. Past conditions of geography, geology, ecology, and climate define the present state and constrain possible future states. Consequently, if you want to understand how any landscape functions now it is useful to understand how it got to its current state; that it is the culmination of a complex set of long-term and interwoven system dynamics. Each strand of this historical tapestry operates on different scales of space and time. Some processes such as tectonics work on a planetary scale over millions of years. Other processes such as weather play out locally over hours. Unravelling the connections and interdependencies between all of these different processes operating on such diverse scales of space and time is an intriguing puzzle. We know a fair amount about the scientific laws governing most of these things. A lot of the processes function within well-known principles of physics, chemistry, and biology. Yet these principles operate and interact in different ways every time. Wherever we look, we see the same processes arranged into different process regimes; the same elements combine to produce novel patterns. To understand the system, we occupy and most importantly why it now stands on the brink of a system transformation we need to first understand how it came to be.

The grand arc of geological history exerts a significant influence on surface processes. Looking at almost any phenomenon on the planet's surface we can see the imprint of geology somewhere: the particular arrangement of landmasses and topography, ocean basins, and bathymetry defines everything from ocean currents to wind-flows, from animal migration paths to human cultural dispersal.

1.3 Extinctions

There is a constant pressure of extinction and evolution within all ecosystems, but chronostratigraphic boundary events are frequently defined by marked shifts in the magnitude of these the changes. These high magnitude events are called mass extinctions, and they do not just clip the leaves of the taxonomic tree (species). They prune away twigs (families), branches (genera), and sometimes entire limbs (orders).

Mass extinctions serve as palpable evidence of major Earth system changes as they are triggered by several possible transformative mechanisms. These include impacts from extraterrestrial objects, episodes of volcanic activity, and internal feedbacks. The relationship between these events can vary in both time and space, but there are some general rules which hold: significant disruptions of Earth system function initiate a predictable set of eight interlinked kill mechanisms:

1. Ocean acidification
2. Toxic metal poisoning
3. Acid rain
4. Ozone damage
5. Increased UV-B radiation
6. Volcanic darkness
7. Cooling
8. Photosynthetic shutdown

1.4 Climate Change

Humans have never lived in a world without ice. When global average temperatures rise above 18 °C the polar ice caps melt. For the 541 million years of the Phanerozoic, more than the entire time in which vertebrates have existed, the world has alternated between "icehouse" and "hothouse" conditions. It has either had polar ice-caps, or it has no polar ice-caps. Over this time there have been three major and six minor icehouse episodes, making up in total around 25% of the total record. For the balance hothouse conditions prevailed (Fig. 1.3).

We are currently in an icehouse period which has so far lasted for around 20 million years. Within this time-frame lies the entire known history of all of the great apes, the entire family Hominidae, all of our hominid ancestors, and of course us. Our entire evolutionary journey following our divergence from other primates has occurred in an icehouse world.

This should give us significant pause. Changes between icehouse and hothouse conditions mark major boundaries in geological time. The lines dividing time slices such as eras, periods, and eons in Earth history each represent a change in system state. Weather patterns change, ocean currents are rearranged, drainage patterns reroute, ecosystems undergo major shifts, organisms become extinct, and new ones evolve. Current anthropogenic climate trends may well deliver us an ice-free Arctic

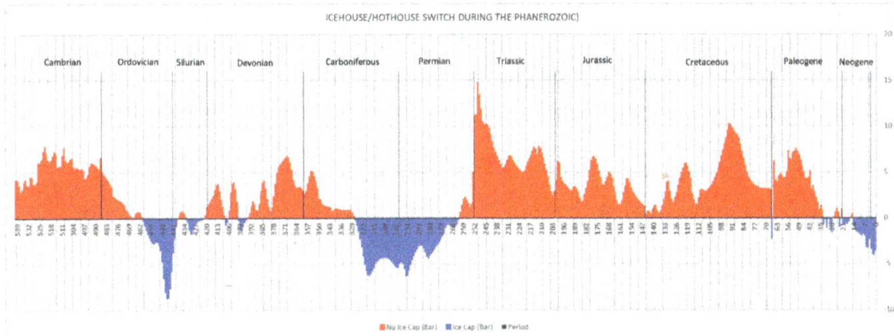

Fig. 1.3 A summary of Phanerozoic climate change reproduced from data presented in (Scotese et al. 2021)

by the end of the twenty-first century. Arctic ice has already declined by at least 40% since 1970. It is possible that by 2035 there will be no Arctic ice in the summer. If this happens models become much less certain about what will happen next. It may be that at this point human actions will become irrelevant. All of what we know about past climate is derived from proxies – those records of past events that were never measured directly, but are preserved in some other tangible form. For example, tree rings tell the story of how many years have passed. And they also, by their width year on year, tell the story of how much rain fell, how hot the summers were, and how cold the winters were. One of the challenges facing a scientist interested in these questions is how to generalize – that is, how to walk from a specific case to a universal trend. Most of the data describing climate in the pat is derived from one of four sources: large pieces of preserved organic material, pollen, sediment, or ice cores. All of these measures give us insight into past conditions by telling us something we can use to deduce, interpolate, or extrapolate from. Absent direct measurement of something, we are forced to rely on indicators.

1.5 The North Atlantic Polar Triangle

The North Atlantic is by geological standards, quite young. Most of its physical characteristics only became recognizable about when the American, African, and Eurasian continents split apart in the early part of the Cenozoic, from about 65 Mya forwards. For a significant proportion of this time there has been some coherence to environmental processes within its bounds. Within the region defined by its bounding landmasses and the equator (Fig. 1.1) there have been coherent patterns of circulation of atmosphere and ocean, interchanges between localized ecosystems, and most recently, cultural and economic connections. Most of the processes reach beyond this boundary. However, as defined, it provides a manageable geographical unit within which to explore and understand the development, and ultimately the predominance of human influences on environmental processes. In part this is also because the Holocene history of the NAPT has been studied in greater detail, and is therefore far better known than most other parts of the world. Thus, neither the tropics, nor the southern hemisphere, nor the pacific have such a rich mine of data from which to draw in building an understanding of their long-term natural histories.

High on the list of reasons for this is that the Northern part of the NAPT is home to a very high proportion of the World's highly developed nations. Therefore, it is also home to a very high proportion of the World's wealth, research institutions, and trained researchers. We have been diligently studying this region since the invention of the formal discipline now known as science. Many scientific "firsts" are associated with the place, so we know it much better than we know other parts of the World.

The first systematic temperature measurements were made in Italy, sponsored by the Medici family, starting in the seventeenth century (Camuffo and Bertolin 2012). The longest continuous set of meteorological observations start in 1781, from the Hohenpeißenberg Observatory, in Bavaria (Deutscher Wetterdienst 1981). In many parts of the less developed world direct meteorological observations have only started within the past century. Long-term instrumental recordings of meteorology are much harder to make away from land. Consequently, the southern hemisphere, which is predominantly water, and the Pacific Ocean, which has only a sparse and scattered land covering, both lack a well- organized network of meteorological observatories. In sum, the high latitude North Atlantic has been better studied, for a number of historical and geographical reasons. It also contains a more accessible record of a longer time period. This situation seems to adds a rather perverse level of privilege to a part of the world that is already significantly privileged. It also serves as a powerful reminder of the importance of supporting scientific research beyond the highly developed world. There are vast areas of the Earth's surface that hold records of past events and can serve to greatly improve our understanding of our planet's history, and therefore to better understand its future. But the quirks of economics, politics and human events have, so far, rendered those records far harder to access.

1.6 The Anthropocene

It has become apparent over the past couple of centuries that some of the changes in the natural environment are the unintended consequence of human action. It seems that almost every realm of natural processes, and every part of the Earth's surface, atmosphere and oceans included, are experiencing changes attributable to human action. There are some obvious markers, such as melting ice-sheets, vanishing forests, and warming oceans. There are also much more subtle hints such as ubiquitous plastics, antibiotic resistant bacteria, and collapsing insect populations. There is a growing sensibility among scientists that humans are probably the primary driver of changing dynamics in natural processes at present. The term used in the Earth sciences for this effect is "forcing factor". This forcing role of human society has led some researchers to propose that a new geological age be recognized, the Anthropocene, which roughly translates as "The Age of Humans". A heated debate continues around the definition of this new age. The date at which humans became the dominant force in natural processes remains in question. The universal stratigraphic marker is defined, but already disputed. The indicator fossil proposed (chickens), but not yet universally accepted (Waters et al. 2014).

There is a good library's-worth of books exploring the role of human action in transforming the physical environment. Almost as soon as formal modern science emerged in the Eighteenth-Century perceptive researchers identified and formalized the role of human action in environmental processes. Alexander von Humboldt and Charles Darwin both commented at length on human impacts as part of their broader scientific writing. The American writer George Perkins Marsh wrote a seminal book "Mand and Nature: Or, Physical Geography as Modified by Human Action", which explored these ideas in much greater detail (Marsh 1894). The book's six chapters: Introductory; Transfer, Modification, and Extirpation of Vegetable and of Animal Species; The Woods; The Waters; The Sands, and; Projected or Possible Geographical Changes by Man anticipate the general structure of most subsequent treaties on the topic.

Despite this rich literature we still struggle to understand how humans fit into Earth history. Leaving aside the absurdities of creationism, the flat Earth movement, and climate change denial, there is still an alarming gap in much of the public discussion around planetary processes. This is evident in common discourse. For example, a recent New York Times column by Maureen Dowd was titled "A.O.C. and the Jurassic Jerks: For the Cave Man President: and his party, clubbing women is not a path to victory" (Dowd 2021). A paper famed for its accuracy and editorial rigor slipped not one but two splendid misdirected metaphors into a single headline. These kinds of gaps in knowledge are extraordinary, as they point to a much broader ignorance of not only Earth history, but also the basic tools and techniques we have used with impressive success over the past couple of hundred years to learn about the planet's history. Every cartoon of a "stone-age man" pursued by a dinosaur, and every dismissive use of "Neanderthal" as an adjective, speaks of ignorance that would be unacceptable for an educated person in almost any other area of human learning.

Dividing human culture into manageable episodes requires some organizing structures. A simple historical subdivision will not serve the purpose of this book, as even within the limited geographical scope I am analyzing there are still some pretty significant disparities in character and history between regions. Likewise, the construction of managed agroecosystems, and the collateral ecosystems emerging beside them continues. However, there have been some fundamental shifts in how these changes are accomplished. The revolutionary transmutational technologies of the early Holocene have been progressively enhanced by an exponentially growing toolbox of other molecular and material manipulations. The transition from bronze to iron is the first step in a journey culminating with the alloys and heavy metals of the Industrial Revolution. Added to these are the huge array of non-metallic synthetic materials, the vastly expanded utilization of biological and mineral resources, and the increase in interconnectedness.

There are many interpretations of available data, and many explanations offered for human behaviors. Some general principles emerge from all of this. It is broadly accepted that there is such a thing as "society". Further, it is broadly accepted that the behavioral characteristics of a society have some collective similarities lumped under the term "culture". My job here is not to relitigate the debates concerning the differences between competing theories of cultural interpretation, but rather to focus upon the shifts in how humans have interacted with their environments. On the most basic level all other interactions people have with what's around them is predicated on an interaction with the physical world. Breathing, drinking, and eating are the necessary precursors to everything else. In this sense economy at its most fundamental can be used to understand how people organize themselves. This useful approach is developed by Ian Simmons (Simmons 1996). One of his central organizing principles is that the relationship between humans and the environment can be modelled in terms of energy use, for which a good proxy is mode of production. The energy principle is widely used in ecology, and he suggests that it serves as a useful surrogate for the extent to which humans transform the world around them.

As understanding of physical processes has improved, as we have learned more of how natural processes function, and as our understanding of Earth history has improved the corpus of literature has grown. A fundamental contribution to the discussion was made in 1956 with the publication of "Man's Role in Changing the Face of the Earth", Edited by William Thomas, Car Sauer, Marston Bates, and Lewis Mumford (Thomas et al. 1956). This weighty tome runs to over 1100 pages, and covers everything from the first human civilizations, through to the prospect of a human-driven demographic and material collapse of Earth systems. This extraordinary range of discussion was echoed again in the 1993 publication "The Earth as Transformed by Human Action" (Turner et al. 1993). This work has a tighter historical focus and is explicitly geared towards human impacts since the Industrial Revolution.

More technical works, focusing on the mechanics of human interaction with the environment from a physical point of view also exist. Prominent among these are Andrew Goudie's "The Human Impact on the Natural Environment" (Goudie 2005), Neil Roberts "The Holcene: An Environmental History" (Roberts 2014), and Ian Simmons "Changing the Face of the Earth" (Simmons 1996). Both undertake to provides a broad overview of the entire Pleistocene "human experiment". Goudie arranges his work topically, organizing it around themes such as "the human impacts on vegetation", and "the human impacts on animals". Simmons uses the axis of human culture to organize things. His general principle is that human history can be divided into economic stages, from hunting, through to industrialization. These stages can be distinguished by technology, energy flow, and social organization. This organization of human history into "Modes of Production" is extremely useful in developing an understanding of how humans interact with their environments. As originally stated by Marx the term includes not only the workaday technologies of productivity, but extends to other areas of cultural life. There are some limitations to the idea. Not unusually for a nineteenth century theory it also contains quite a few empirical errors, and there are some failings in the model's predictions: we are still awaiting the dictatorship of the proletariat.

This noted, the similarities between, for example, the way fishing, hunting and gathering cultures interact with the natural environment, wherever and whenever they existed, are greater than the dissimilarities in some key ways. There are some simple metrics that can be used to explore this idea. Low population densities, low fertility rates, and low energy consumption characterize hunting cultures from tropical rain forests, through deserts, temperate woodlands, tundra, and Arctic landscapes. Whether coastal, mountainous, or continental, there are common structures of social organization, common technological strategies, and common interactions with the physical environment. Likewise, pre-imperial farming societies tend to have a prevalence of common characteristics. Although the range of physical environments in which farming is found is somewhat smaller, within this range similarities exist. The manipulation of physical environments to manufacture and maintain an agroecosystem that provides for basic human nutritional needs, the domestication of animals for food and for labor, organization of production spaces around particular economic functions, hierarchical social structures that manage labor.

Over the past century there has been a growing sense among Earth and environmental scientists that humans are a major force in determining the ways in which natural processes operate. This idea descends from the somewhat older notion that the human historical journey has been one of every growing dominion over nature. In the modern industrial world this idea of dominion over nature is embedded in an overwhelming majority of our narratives. Within the Judeo-Christian tradition our sacred texts from Genesis forwards describe an entire universe "Given unto Man", who was created in God's image, and who then given naming rights over birds, beasts and fowls. Our national identities are described by some of the more excitable among us in terms of "blood and soil". In this story we struggle against and triumph over nature. Medieval monks settled in remote and hostile places and eked out a living as a form of prayer. Europeans bravely sailed to wild lands inhabited by savage peoples and tamed them both. We mastered fire, made the land fruitful, tamed animals, controlled the waters, and brought nature to heel. In the modern industrial world classical capitalist and communist ideologies both regard "mother nature" as subjugated to the will, the strength, and the knowledge of an ascendant humanity.

As noted earlier, geological divisions of time are defined by changes in process. Only very rarely do many processes change simultaneously over large areas of the Earth. Far more frequently changes are incremental and piecemeal. Although they may appear sudden in the geological timeframe, they may take thousands or millions of years to affect the whole world. Thus, even the coveted "golden spike" of chronostratigraphy may represent long periods of time and may be hard to identify globally. One of the fundamental underlying principles of geological science is Uniformitarianism. This is often characterized by the phrase "the present is the key to the past". Simply stated this is the idea that processes have always worked the same way. If a pebble has been rounded by bouncing along the bed of a stream in flowing water, then other similarly rounded pebbles were at some time also bounced along in flowing water. On one level this is a very useful idea since it is based on the not unreasonable notion that the physical laws of the universe remain the same through time (whether we have correctly described them or not). The term Uniformitarianism was first proposed by the enlightenment-era scientist William Whewell – who also conveniently coined the term "scientist" (Heilbron 2002). The idea was put to good use by the pioneering geologists James Hutton (who used the concept before it was named), Charles Lyall, and John Playfair in various forms to embark on interpreting the complexities of Earth history by decoding the stratigraphic record, and to simultaneously counter creationist thinking.

That scientific laws hold throughout the universe remains a central tenet of science. However, things have become a bit more nuanced in the past two hundred or so years. For a start there are a great many more scientific laws now, meaning that the central tenet of uniformitarianism can be used to explain a much broader array of phenomena. This means that the current consistent set of physical laws can now explain natural phenomena that would have been inconsistent and anomalous under the old rules. One logical outcome of this is that when scientists are confronted with new inconsistencies and anomalies, they work to update the laws, rather than abandon the principle of scientific laws altogether. A more serious deficiency of classical Uniformitarianism as expounded by Hutton, Lyell, and Playfair is proposed by Stephen Gould in his book "Time's arrow, time's cycle" (Gould 1987). Gould argues that Uniformitarian-based explanations are derived from a core principle that things happened in the past the way they happen now. Using this notion can predict, among other things, predict the future. As a paleontologist Gould notes that as far as living things are concerned, the past was very different from the present, and the future will be very different from the past. His argument is summed up in his book's title – Earth history is not a cycle, it has direction. Change is linear, and we cannot return to past states. Rather we inexorably move forwards to new states. The characteristics of each successive time period, however defined, are determined by the conditions that preceded it. This idea works pretty well in evolutionary biology. There is no possibility of dinosaurs re-evolving (sorry, kids). However, it is possible to imagine a set of conditions in which a dinosaur like organism might evolve from a currently existing species, were the correct sequence of environmental changes to emerge over a sufficient period of time to which that was an effective evolutionary response.

The directionality of Earth history is, in fact, foundational in much of our understanding of the physical universe. The divisions of geological time are premised upon the notion that there are fundamental shifts in the broad organization of processes, life, and the arrangement of the geographical features of the surface through time. The Proterozoic is distinguished from the Phanerozoic because a whole bunch of things worked very differently after the division. However, there are also equally well-established foundational notions which presume cyclicity in some processes. For example, we have the life cycle, the hydrological cycle, the rock cycle, and Milankovitch cycles. Gould argues that, rather than discarding these useful intellectual tools, we consider the impact of metaphor on our understanding of natural processes. He proposes the idea of a spiral, rather than a cycle, as a more effective metaphor for Earth history. This allows for both directionality and a return to similar states, constrained by the universality of physical laws (Gould 1987). This in turn presents a few difficulties, as it does not give us room to consider dramatic change events such as meteoric intervention in historical events – those (thankfully) rare moments in Earth history when everything suddenly changes in unprecedented ways.

In the context of this book it is this notion of change – when “business as usual” ends, and a new system state is established – that is central to understanding how things work. When is a change big enough to be a significant change? Is such a change invariably catastrophic? Within which timeframe or over what area must such a change occur to be considered significant? The boundaries between chronostratigraphic units, large and small, mark a transition between system states. In a few cases some cases these shifts are the result of a single dramatic event. The most commonly cited example of such a shift is the Chicxulub meteor strike. In a single catastrophic event, the entire Earth surface system was rendered unstable. Most other changes are far slower and occur over longer periods of time. Typically, a cumulative quantitative change ultimately precipitates a qualitative change. This is frequently driven by multiple nesting and interrelated positive feedback loops. All available evidence now indicates that we are at a crucial moment of change: the Anthropocene is most probably a boundary event. How big the change, and how protracted remains to be seen.

1.7 Structure of the Book

This book is organized into six chapters. This introduction lays out the parameters of the discussion, provides some background information, and introduces some key technical terminology. The next chapter covers the whole of Earth history from the formation of the planet, around 4.6 billion years ago, through to the end of the Last Glacial Maximum (LGM) covering this vast sweep of time serves a couple of purposes. First, it outlines the depth of time comprising geological history, and gives some perspective to the place in that history of the past 11,700 years. What may seem to us to be monumental events are, in the grand scheme of things, of negligible significance. We are not, it turns out, really that big a deal. In the context of all geological history it is possible to understand the relative significance and insignificance of changes currently underway around us. Secondly, this chapter outlines the events that set the stage for the Holocene to unfold. As this book has a spatial as well as a temporal focus it is important to get a sense of how the current geography of the North Atlantic came to be, and to understand what is currently within the study region. This includes the arrangement of land and water, the topography, the distribution of flora and fauna, and the surface characteristics that influence climate, hydrology and oceanography. In this respect, within the context of all of geological history the Earth’s current geography is simply a few arbitrary frames from a long-running movie. Although of huge significance for the day-to-day operation of processes right now, we know that things were different before, and will change again.

The next three chapters cover the three major chronostratigraphic divisions of the Holocene, the Greenlandian, the Northgrippian, and the Meghalayan. These are defined by physical traces in the landscape and represent major shifts in the operation of environmental processes. These three divisions of the Holocene are not determined by human historical stages, although there are some parallels between how humans lived and these ages. The climate trend that has defined these three ages is outlined by Marcott et al. (Fig. 1.4). Although there are significant regional variations, and temporal shifts, the general trend in the global average is clear. A rising arm, a gently declining plateau, a falling arm, and a terminal spike. This curve has, to the best of my knowledge, not been formally named. To my eye it resembles the Arabic letter ح (pronounced "ha") rotated 90° clockwise. Absent any other generally accepted title I shall call it the ح-curve. One notable characteristic of this curve is that it contains two striking deviations from the prevailing form. One of these, at around 8200 BP is known as the 8.2 K event, and it marks the end of the Greenlandian Age. The other, of comparable magnitude, is the terminal spike. One of the central arguments of this book is that we are currently at the end of an age, or possibly of an entire period. As I have suggested earlier, I am uncomfortable with defining a new period called the Anthropocene for a number of empirical and theoretical reasons. I feel more comfortable discussing the end of an age. I shall argue throughout this book that the terminal spike, and its associated causes and effects mark the end of the Holocene. I shall leave it to future geologists to name what comes next, as only in retrospect will its prevailing characteristics become apparent.

The Greenlandian runs from 11,700–8200 BP and starting with the end of the LGM. It was a time of consistent warming, with global average temperatures going up by about half a degree centigrade until a catastrophic cooling event around 8200 years ago, the 8.2 k event. This was probably associated with the final collapse of the Laurentide Ice Sheet, and a series of significant tsunami events. If there is a historical equivalent of the Biblical flood, this was it! At the beginning of the Greenlandian the World's geography was very different from its current configuration. The first 3500 years of the Holocene saw a significant change in sea levels and ice cover, which translated into significant changes in the amount of habitable land, and the land bridges connecting a number of coastal regions and what are now offshore islands. Similarly, ice bridges and barriers changed the patterns of possible communication between different parts of the Holocene world, closing some routes, and opening others. There were a series of megafaunal extinction events associated with the end of the LGM, and the arrival of humans in previously unpeopled landscapes, but these had been underway during the Ica Age as well. Wherever people showed up, the big animals died. Humans started a number of significant shifts in behavior during this time. Although dogs had been domesticated during the LGM, there is no other evidence of domestication in the archaeological record until the Greenlandian. During this period the first systematic efforts to change human economy away from hunting and gathering occurred. The first evidence of ceramics and the first evidence of smelting also occurred. Within the classical archaeological

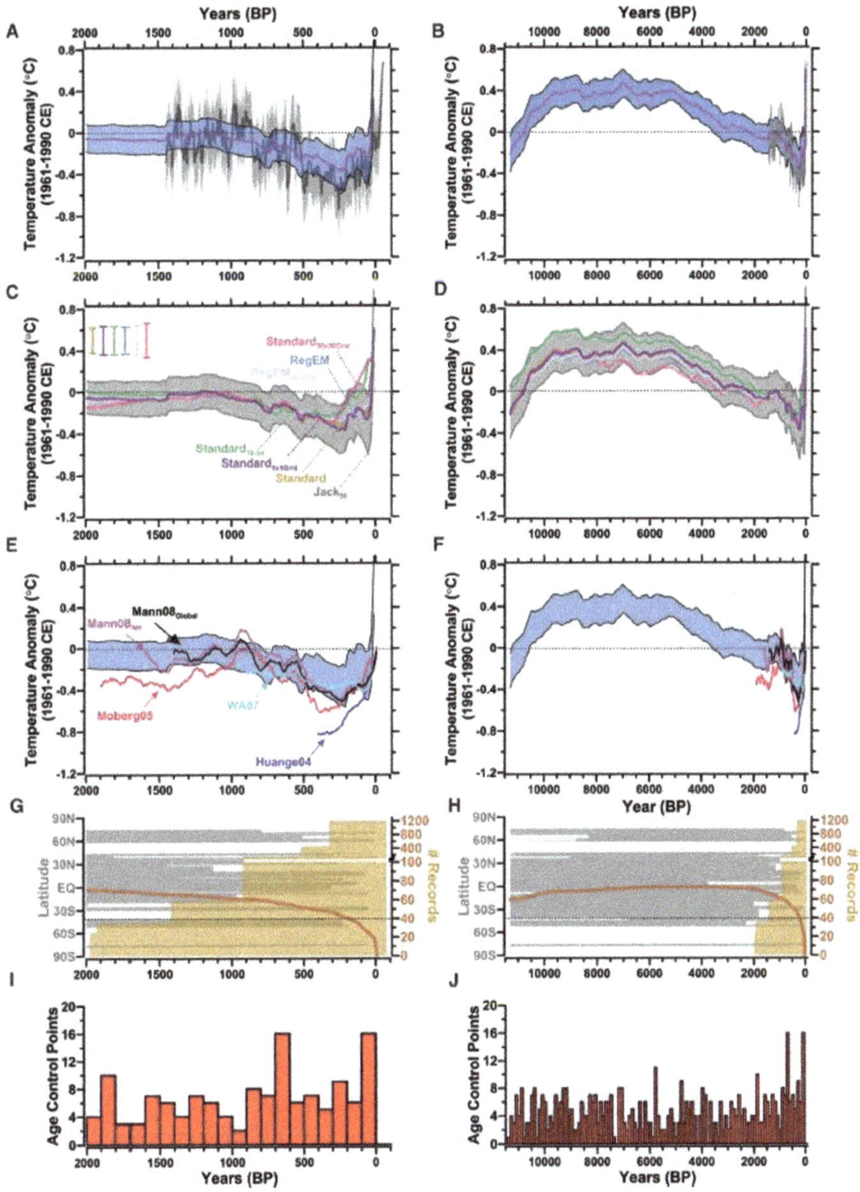

Fig. 1.4 Comparison of different methods and reconstructions of global and hemispheric temperature anomalies. (Marcott et al. 2013)

model this time is typically associated with the Mesolithic, a time of technological transition. The venerable Paleolithic traditions of toolmaking were replaced by new approaches to survival, and also some striking new cultural practices. These included the establishment of permanent settlements, and the construction of complex ceremonial structures. These practices were not universal, nor were they consistent through time. The new adaptations spread outwards from a few centers of innovation, gradually displacing the old ways.

The Northgrippian runs from 8200–4200 BP. It was a period that encompasses the thermal maximum of the Holocene and was characterized by gradual cooling. It ended with the 4.2 k event, a global drought. The world's geography stabilized into its current configuration. Human cultures underwent an extraordinary transformation during this period. The nascent domestication and farming practices of the Greenlandian developed into full-scale farming cultures. Humans began to construct complex agroecosystems, that seem to have superseded hunting and gathering lifeways wherever the two came into contact. Ceramics and metallurgy both underwent significant evolutions. Although not universally adopted – for example in the Americas early localized copper smelting and toolmaking lasted for a while and then people returned to using stone tools – these transmutational technologies were gateways for a whole host of other undertakings. They served as force multipliers – you can cut trees faster with metal tools, even copper. They also opened other options, including cooking, fermentation, food storage, and food preservation.

During the Meghalayan, which runs from 4200 to the present, an extraordinary shift occurred. This timespan covers most of recorded human history, and we can trace direct cultural, technological, and genetic links into this past. The broad climate trend of this time was a prevailing decline, followed by a dramatic spike. This spike is unusual in the Holocene, as is simultaneous, and global, in ways that other changes within the period have not been (with the possible exception of the 8.2 k event). The Meghalayan is also notable for the general trend of acceleration in all things human. During its 4200-year span humans built most of the great NAPT empires. This covers a huge range of cultural and geographical space, from the classical Mediterranean civilizations through pre-Columbian meso-America, Medieval Europe, the emergence of the post-Medieval global trading empires, and the development of the Industrial world. For most of the Meghalayan there has been a concentric expansion of new modes of production outwards from core regions, with rival modes of production pushed to the peripheries of the region. In 1691, a little over three hundred years ago, the NAPT hosted almost the entire range of Holocene human adaptations. The American Arctic, Greenland, the deserts, and the tropical forest regions were occupied by hunter-gatherers, some of whom used neither ceramics nor metals, and used only a single domesticated species: dogs. Simultaneously there were non-imperial farmers in the Celtic fringes of Europe, the Caribbean, the American eastern seaboard, and the African coastal tropics. Imperial agriculture thrived in the Americas, and in West Africa. At the same time the first technological, economic, and cultural seeds of urban industrial societies started to develop in Europe.

A central part of my argument is that we are now at the end of the Holocene, which is therefore also the end of the Meghalayan. I argue that the declaration of the dawning of a new age, the Anthropocene, is illogical. It is impossible to accurately define a geological time slice before it has occurred. Rather, we are currently witnessing the boundary event. The sixth chapter is, then an overview of the current situation. I have argued that the Meghalayan "spike" marks the end of an age. We are currently experiencing a dramatic change in process regimes within and beyond the NAPT. Barring some dramatic shift in driving forces, such as an asteroid-strike, or a collapse in human population we can predict with some accuracy the general trajectory of change. It is not necessary to advocate for any particular course of action in this section, as the outcomes are fairly easy to anticipate. For example, most terrestrial and marine megafauna are already well on the way to extinction. It is highly unlikely that any of the large whale species will survive long into the twenty-second century. Likewise, it is highly unlikely that tropical rainforests will exist in any recognizable form in a hundred years. Species diversity of both plants and animals is already dramatically reduced. Human populations are continuing to expand rapidly. Climate trends seem set, with global temperatures consistently increasing year-by-year. There are a well-documented and predictable set of outcomes for the projected changes. These include sea-level rise, tropical desertification, and species migration. Other outcomes can be hypothesized but are not so well documented. These include such things as habitat changes, evolution of surviving species, and human cultural change.

Earth history indicates that changes resulting from major shifts in processes are complex. The patterns produced by a significant shift in a forcing factor propagate outwards and resonate at many frequencies. A single species extinction can have immediate effects on its predators, prey, and parasites. It can also have longer term impacts on other ecosystems separated in time and space from its normal habitat. One example of this is the phenomenon of whale-fall, where fallen whale carcasses in deep ocean plains provide decades, and in some cases centuries, of habitat for specialized abyssal scavengers. These islands of nutrition in otherwise relatively barren environments provide oases for some animals moving between widely separated locations in a challenging environment. It may be centuries after the death of the last whale before the last of these dependent detritivores is driven to extinction or adaptation (Smith et al. 2015). Soil erosion triggered by mountain deforestation has a series of impacts that may take centuries to play out. Initial slope degradation is followed by channel aggradation. As the bolus of sediment is transported downstream, often over long time periods, is impacts each region it passes through in sequence. Final deposition in an alluvial fan or in a delta can take centuries or millennia. Other system changes can variously enhance, offset, or overwhelm the effects of the initial erosion (Bampton 1993).

References

Alvarez W (2008) T. rex and the Crater of Doom. Princeton

Bampton M (1993) Anthropological influences on the geomorphology of the Ebro Delta. PhD Thesis. Clark University. Worcester, MA

Camuffo D, Bertolin C (2012) The earliest temperature observations in the world: the Medici network (1654–1670). Climatic Change 111(2):335–363

Crutzen PJ, Stoermer EF (2000) The Anthropocene. IGBP newsletter, 41. Royal Swedish Academy of Sciences, Stockholm

Deutscher Wetterdienst (ed) (1981) 200 Jahre meteorologische Beobachtungen auf dem Hohenpeißenberg 1781–1980. Offenbach am Main

Dowd M (2021) A.O.C. and the Jurassic Jerks: for the Cave Man President and his party, clubbing women is not a path to victory. New York Time July 25, 2021

Goudie A (2005) The human impact on the natural environment. Blackwell, Oxford

Gould SJ (1987) Time's arrow, time's cycle: myth and metaphor in the discovery of geological time. Harvard University Press, Cambridge

Heilbron J (2002) Coming to terms. Nature 415:585

Jackson PW (2006) The chronologers' quest: the search for the age of the earth. Cambridge University Press

Marcott SA, Shakun JD, Clark PU, Mix AC (2013) A reconstruction of regional and global temperature for the past 11,300 years. Science 339(6124):1198–1201

Marsh GP (1894) Man and nature, 2003rd edn. University of Washington Press, Seattle

Pielou EC (2008) After the ice age. University of Chicago Press, Chicago

Pinker S (2021) Rationality: what it is, why it seems scarce, why it matters. Viking, London

Roberts N (2014) The Holocene: an environmental history, 3rd edn. Wiley-Blackwell, London

Scotese CR, Song H, Mills BJ, van der Meer DG (2021) Phanerozoic paleotemperatures: the earth's changing climate during the last 540 million years. Earth Sci Rev 11:103503

Simmons IG (1996) Changing the face of the earth: culture, environment, history, 2nd edn. Wiley-Blackwell, London

Smith CR, Glover AG, Treude T, Higgs ND, Amon DJ (2015) Whale-fall ecosystems: recent insights into ecology, paleoecology, and evolution. Annu Rev Mar Sci 7:571–596

Thomas WL Jr, Sauer C, Bates M, Mumford L (1956) Man's role in changing the face of the earth. University of Chicago Press, Chicago

Turner BL, Clark WC, Kates RW, Richards JF, Mathews JT, Meyer WB (eds) (1993) The earth as transformed by human action: global and regional changes in the biosphere over the past 300 years. Cambridge University Press

Waters CN, Zalasiewicz JA, Williams M, Ellis MA, Snelling AM (2014) A stratigraphical basis for the Anthropocene? Geol Soc Lond, Spec Publ 395(1):1–21

Chapter 2
Before the Holocene

From 4.6 Billion Years Ago to 11,700 BP

Abstract Earth history, from the formation of the planet to the end of the last glacial maximum (LGM), emphasizing boundary events. Perspectives on geological time. The distinction between magnitude and frequency in Earth surface processes.

Keywords Geological column · Hadean · Archean · Proterozoic · Ceonzoic

2.1 Introduction

This chapter is a brief overview of Earth history before the Holocene. Readers familiar with this topic may have some differences of opinion on the weight I have given to, and the interpretation I have made of, some topics. There are a lot of active debates about the details of Earth history, and there are many unresolved questions remaining. I have focused on the parts of this history that pertain to the formation and current configuration of the study region. I have also placed particular emphasis on the boundaries that separate eons, periods, and epochs as these divisions in the planet's history are an important part of my overall argument that we are currently witnessing a boundary event. In general, I have tried to stick to the broadly accepted timelines, and accounts of causes, effects, and processes. Given a temporal scope of 4.6 billion years and the constraints of chapter length I have glossed over many details and ignored active debates. I have relied on a combination of standard texts, recently published survey papers, primary research papers, and some popular accounts written for lay audiences, in an effort to provide an account that is simultaneously current, accessible, and accurate. Historical detail increases as we approach the present for two reasons: first, because the geological record is much richer for more recent events, and secondly; because rates of change, particularly in organic systems at all scales, has accelerated over time. As ecosystems, became more complex, and were comprised of a greater number and diversity of organisms,

M. Bampton, *The North Atlantic Polar Triangle*, Springer Polar Sciences,
https://doi.org/10.1007/978-3-031-27264-6_2

they tended to change more rapidly. These rapid changes in ecosystems had a knock-on effect, accelerating changes in many inorganic systems, such as the atmosphere and oceans.

The North Atlantic is by geological standards, quite young. Most of its physical characteristics only became recognizable about 65 Mya, when the American, African, and Eurasian continents split apart. But understanding it requires us to understand its origins. Packing almost the whole of Earth's history into a single 30-page chapter to provide context for a more detailed discussion of a tiny slice of very recent time is an ambitious undertaking. However, I am not attempting to provide a comprehensive review of historical geology. There are excellent specialist works that address that topic. I have leant heavily on several of them to write this chapter. The chapter provides context for the more detailed discussion of the Holocene that comes later. To fit most of 4.6 Gya into one chapter I have summarized complicated processes and major events in a few sentences or paragraphs. I have skipped many things completely, and I have ignored ongoing debates. Where possible I have provided references to more complete explorations of important ideas. In some cases, there are seminal books on key topics. A significant proportion of the literature is primarily topical. Several of these works stand out as particularly useful for giving a broad perspective on a wide variety of materials. David Beerling's "The Emerald Planet: How Plants Changed Earth's History" (Beerling 2017) is an excellent summary of the entire history of plant lie on the planet, from the Proterozoic to the Present. A more detailed view of the Precambrian story is available in Andrew Knoll's "Life on a young planet: the first three billion years of evolution on Earth." (Knoll 2021) fills in a lot of detail about pre-vascular plant life. It also details the early impacts of plant-life on Earth's climate.

There are also excellent books written for general audiences exploring key ideas. In some cases, I have referred to original research papers, or survey papers published in the technical literature. All of these sources provide valuable insights into how the earth evolved, and how we got to where we were at the beginning of the Holocene. This chapter functions as a concise encyclopedia article on Earth history, from the formation of the planet to the end of the Last Glacial Maxima (LGM).

The chapter has four sections: the Precambrian, the Paleozoic, the Mesozoic, and the Cenozoic. The Precambrian, earliest and longest part of Earth's history, can be defined roughly as the time before complex life evolved. Atmosphere and ocean formed during the first eon of the Precambrian, the Hadean. The first living things evolved in its second eon, the Archean. In its final eon, the Proterozoic, saw the first animals appear. The Paleozoic, the Mesozoic, and the Cenozoic, are respectively the times of ancient, middle, and recent life, and are the subdivisions of the Earth's fourth (and current) eon, the Phanerozoic. Each of the chapter's four sections is divided into several subsections, based on some of the smaller named slices of time. As we get closer to the present, the time slices get smaller to address in greater detail the changes occurring in a world which is progressively getting more and more similar to the one we currently inhabit.

2.2 Precambrian Eons

The 4.1 billion years before the Cambrian, from 4.6–541 Ga, are lumped together into the time known as the Precambrian. This comprises the majority of the planet's history, and it is often represented on graphics of the geological time scale as a small dark block at the base of the figure, with a ragged lower edge to illustrate its disappearance into the mists of the past. Everything after the Precambrian is the Phanerozoic. The International Commission on Stratigraphy solves the graphical problem by using different scales for the Precambrian, and subsequent times (Fig. 1.2).

Although its extreme antiquity makes it hard to study, we have increasingly good models of the Earth during this time. In this section I'll echo the ICS figure and compress each of its three eons into short discussions that focuses on distinguishing characteristics. The Hadean, the Archean, and the Proterozoic, could respectively be summarized as the start of geology, the start of geography, and the start of biology, as by the end of each one of these major systems was in place.

In the beginning (that is, before about 4.6 billion years ago) the solar nebula, a spinning cloud of dust and gas, occupied the space where our solar system now sits. This cloud differentiated into several clumps, with the largest and hottest of these, the Sun, at its center. The majority of the remaining materials consolidated into a series of lesser objects in concentric orbits around the Sun. The third of these was the Earth. The newly aggregated material cooled and compressed under the force of its own gravity, forming a dense mass with three concentric components, a semi-solid core, a liquid mantle, and a solid crust. Apart from its position in the solar system, its basic mineral composition, and the three-part structure, the early Earth was in most other respects fundamentally different from its current form.

2.2.1 *Hadean*

The 0.7 ka of the Hadean, from 4.6–4 Ga, are the most obscure time in Earth's history. For most of the it the crust was probably homogeneous, rather than differentiated into two kinds of material as it has been ever since (Drabon et al. 2021). There were no large chunks of low-density continental crust sitting on plates of denser oceanic material. Instead, the relatively homogenous surface alternated between "stagnant lid" and "overturn" phases. Periodically, in response to the pressure of rising magma plumes, or from external impacts, portions of the outer layer of the planet flipped over in a "crust-mantle overturn event". Even in the stolid language of geology this sounds impressive. In such a dynamic environment, only very robust materials survived. These were mostly zircon crystals, and they constitute the bulk of the geological record of the Hadean.

Around 4.4 billion years ago, an atmosphere and water started to accumulate on the surface. The first atmosphere was devoid of free oxygen, and it remained so for about 2 billion years. It was mostly made of ammonia, methane, nitrogen, carbon dioxide, and water vapor. The presence of abundant liquid surface water on Earth is, as far as we know, unique in the Galaxy, and possibly even in the universe. The first oceans were superheated water that only remained liquid because of an atmospheric pressure almost 30 times its current level. Both atmosphere and ocean probably originated from some combination of volcanic outgassing, and materials harvested from incoming or passing comets and meteorites.

The origins of contemporary tectonic structure are still debated. Recent work suggests that episodes of bombardment by meteorites in the middle of the eon may have triggered a shift towards the differentiated crust we currently see (Mauyama et al. 2018). At least one such cataclysm established several of the planet's defining characteristics. About 4.5 billion years ago, a large celestial body, sometimes called Theia, struck the Earth, chipping off enough material to form the moon. The impact also knocked the axis off kilter by about 23.5°, creating the conditions for the subsequent pattern of seasons, the distribution of climatic regions across the Earth's surface, and the patterns of the tides.

Around 4 billion years ago, the crust stabilized and started to form into cratons, the foundational structures of continental plates. This process, and the presence of atmosphere and ocean, marks the end of the Hadean, and a qualitative change in the nature of the geological record, in essence the beginning of the modern geological record (Kamber 2015).

2.2.2 Archean

The Archean, from 4 Ga–2.5 Ga, starts with recognizable geology in the form of the Isua Greenstone Belt, in Greenland. This formation fits into the modern geological schema, as a heavily metamorphosed combination of volcanic and sedimentary rocks. It also shows credible traces of organic activity. This means that early in the Archaean the four cornerstones of geography were present: differentiated crust, oceans, atmosphere, and life. While each of these things was extremely different during this distant past, there were some significant and important similarities with their current forms. A number of processes have worked in comparable ways from the Archean to the present day.

The differentiated crust responded to the dynamics of plate tectonics with continents carried along on moving plates of oceanic crust. Fractional differentiation, the selective crystallization of minerals with different melting points, separated materials of different densities from one another. The lower-density continents accumulated on top of more dense oceanic crust. Convection within the mantle moved oceanic crust around on the surface. Denser materials were forced downwards, back into the mantle, at subduction zones. Less dense materials remained afloat. Consequently, continents, which now cover about a third of the Earth's surface, are usually much older than the material they rest on. Since the Archean new ocean crust has been constantly emerging from the mantle, forced upwards through mid-oceanic ridges. It returns to the Earth's interior at subduction zones, where plates collide. During its (geologically) rapid cycle oceanic crust carries continents along with it like so much flotsam, constantly rearranging the pieces by bumping them against one another, joining them together, and breaking them apart. The majority of the oldest rocks are gone, having long ago melted back into the mantle, or been transformed beyond recognition by weathering or metamorphosis. A few slivers, almost always continental crust, remain.

Although the majority of oceanic crust formed less than 200 Mya, there are a few places where much older materials have survived. In Western Australia there is oceanic crust dating to 3.24 billion years ago. Such quirky fragments contain traces of oxygen isotopes, suggesting there may have been a time during the Archean in which a shallow ocean submerged a large proportion of the Earth's surface (Johnson and Wing 2020). These extensive oceans, far more hospitable than their predecessor in the Hadean, provided the conditions for the establishment of the first ecosystems. It still lacked dissolved oxygen. And it was still hot, around 70 °C. But it was steadily cooling (Garcia et al. 2017).

The early Archean atmosphere was also oxygen-poor, and was still mostly composed of methane, carbon dioxide, and hydrogen sulfide. Nevertheless, the presence of an atmosphere and an ocean created the conditions for two important processes: the poleward circulation of solar energy from the tropics, and the conditions for the evolution of complex ecosystems.

While the general principles of the emergence of life on Earth are known – the creation of "primordial soup" was memorably demonstrated by Julia Childs – many of the details are still debated (https://massasoit.instructure.com/courses/346438/pages/video-the-primordial-soup-with-julia-child). The subsequent relationship between these first ecosystems and the changing chemistry of the oceans is complex. However, some details are generally agreed upon. The bacteria and archaea which formed the first known life, in the form of algal mats, had probably developed by 3.7 Ga. These initially evolved to survive in the non-oxygenated oceans by metabolizing carbon dioxide to produce methane. Over time, some developed the capacity to metabolize carbon dioxide and produce oxygen as a byproduct of photosynthesis. This oxygenic photosynthesis drove a positive feedback loop that incrementally increased the oxygen content of the seas, and also drove their predecessor anoxic organisms to the fringes of the habitable world. Although at this stage only the base of the trophic pyramid existed, the trend of living things to evolve into increasingly complex organisms, and to create increasingly complex ecosystems was initiated. Andrew Knoll summarizes the situation like this:

> Geologically it appears to have been a world of familiar processes, but not-so-familiar patterns. Continents began to form at least 4.2 billion years ago, and chemical details of rocks from Barberton Warrawoona, and other old terrains suggest that large volumes of continental crust had formed by the time they were deposited. Little of these early continents remains, however, implying they were recycled back into the mantle more easily than they are today. Three and a half billion years ago, plate tectonics had already begun to pattern our planetary surface, but Earth's upper mantle appears to have been hotter, the basaltic crust beneath the oceans thicker, and perhaps, the continents smaller and less stable. (Knoll 2021:67)

The change in ecology over the course of the Archean, culminating in the "Great Oxygenation Event" (GOE), straddling the Archean-Proterozoic boundary. Between 3.85 Gya and 1.85 Gya oxygen levels first in the ocean, and subsequently in the atmosphere, rose. This process had several stages. Between 2.45 Gya and 1.85 Gya oxygen rose to between 2% and 4% of current values, mostly in the oceans. This level stabilized until around 0.85 Ga, when surplus ocean oxygen started to outgas into the atmosphere. From this point forwards atmospheric oxygen levels rose towards their present level of 1 atmosphere (spectacularly spiking around 300 Mya but that part of the story comes later). There are still questions about what started this process, but there is consensus that once underway it was driven by photosynthesis.

Once the Earth's surface environments became rich in highly reactive oxygen three important changes occurred. First, the Earth cooled rapidly because the changed composition of the atmosphere allowed more of the sun's energy to escape back into space. Secondly, a whole suite of inorganic chemical reactions became possible, as all sorts of materials could now oxidize. Thirdly, larger and more complex organisms evolved as their more rapid oxygenated metabolisms allowed them to exploit a wider array of nutrients.

The upper boundary of the Archean is defined by the widespread extinction of anoxic species and the matching rapid spread of the rival oxygenic photosynthesizers. The new surface environment had abundant oxygen, novel materials, and vacant ecological niches. This created the conditions in which more complex oxygen-dependent eukaryotic organisms evolved.

2.2.3 *Proterozoic*

The Proterozoic (2.5 Ga–541 Mya) was the longest of the Precambrian eons, lasting around two billion years. During this time the continuing increase in oceanic and atmospheric oxygen allowed for the emergence of complex ecosystems. These were still confined to the oceans – if there was any terrestrial life during this time it was pretty limited. Yet there are Proterozoic rocks on every continent, and recognizable components of every contemporary continent originated during this time – though they have been rearranged quite a bit in the intervening years.

The first reliable map of global geography is from the Proterozoic, with the sequential formation of three supercontinents. First Nuna (sometimes called Columbia or Hudsonland), formed from 1.8 to 1.5 Gya. Then between 1.1 Gya and 900 Mya almost all of the Earth's continental crust was again amassed into a single structure called Rodinia. Between 750 and 633 Mya this broke apart, but by the end of the Proterozoic, between 600 and 400 Mya, its fragments reassembled over the South Pole into another supercontinent, Pannotia (Nance and Murphy 2019). All three supercontinents were so massive that they altered the circulation patterns of Earth's interior, and the character of its rotation. They, and their accompanying super oceans, also allowed for relatively uninterrupted circulation of atmosphere and water. As they broke apart and chunks of continental crust spiralled outwards across the Earth's surface shallow oceans formed between them. One of these, Iapetus, sometimes called the "paleo-Atlantic", sat between the palaeocontinents Laurentia, Baltica and Avalonia (Fig. 2.1).

The Proterozoic experienced at least four lasting cold periods. Between 2.4 Gya and 2.1 Gya two glaciation events occurred, the Huronian and the Makganyene. These are contemporaneous with at least part of the Great Oxygenation Event, and the Huronian also serves as a marker for the beginning of the eon. The second two comprise the Cryogenian period, amd are known as the Sturtian from 720 to 660Mya, and the Marinoan, which ended around 635 Mya close to the end of the Proterozoic. At times during some of the glacial episodes ice extended far beyond the poles into much lower latitudes, possibly from pole to pole, creating a "snowball earth". Whether or not the Earth was ever completely frozen, the Precambrian ice ages certainly lowered sea levels, by freezing much of the Earth's liquid water. If both continents and oceans were largely frozen over only a few isolated spots around volcanoes and mid oceanic ridges would have remained ice-free. The global-scale loss of habitat would have triggered a mass extinction, with the few warm places providing the only refuges for surviving organisms. Although presenting serious

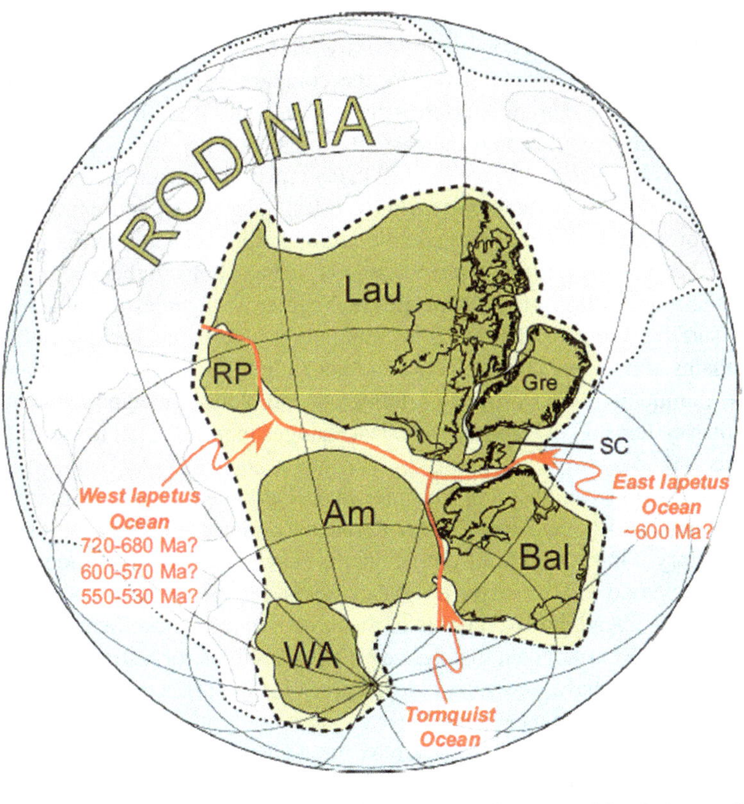

Fig. 2.1 The paleogeography of Rodinia, and the formation of Iapetus. (After Robert et al. 2021)

challenges for these early life-forms the glacial episodes ultimately created conditions in which life prospered and diversified. The ice sheets also ground a significant amount of surface rock into fine sediment and released a great many mineral nutrients into the oceans. As with subsequent extinctions the ecological spaces opened by widespread extinctions provided opportunities for rapid evolution when conditions improved.

The prevailing ecological trend of the Proterozoic was towards increasingly complex and more numerous organisms inhabiting increasingly complex and extensive ecosystems. This was driven in part by the increasing prevalence of photosynthetic food-webs, in part by the increasing diversity of habitats made possible as the Proterozoic supercontinents broke apart and reformed, and in part by climate-driven shifts in sea-level. The end of the Proterozoic saw an increase in available nutrients in an increasingly oxygenated environment, which set the stage for the Cambrian explosion. Or to put it another way, the "invention" of biology.

2.3 Phanerozoic

The Phanerozoic is the shortest of the eons as it has lasted only 541 million years so far. It is also the one we know the best, as we inhabit its most recent epoch, the Holocene. It has been characterized by the comparatively rapid development of complex and diverse marine and terrestrial ecosystems. It is also a time in which those ecosystems have become increasingly involved with the functioning of the Earth's non-living systems. There are still many independent forces which influence Earth surface processes, such as tectonics, tides, solar wind, and extra-terrestrial objects. However, during the Phanerozoic internal feedbacks within the Earth system have become increasingly important. Oxygenic photosynthesis continues to change the composition of the atmosphere. Vegetation changes the albedo of the planets surface. Organic sediments accumulate, solidify into rock, weather, and erode. Plants and animals rearrange the carbon budget. Most recently humans have become a uniquely effective force of deliberate and planed change in almost all of these other systems. I will explore these last changes in more detail in the following chapters.

In the following sections of this chapter, covering the Palaeozoic, Mesozoic, and Cenozoic eras, I will emphasise three patterns that define our understanding of geological history. First, the role of climate shifts between an Earth without ice-caps (hothouse) and one with ice-caps (icehouse). Secondly, the role of mass extinctions, however defined. And thirdly, the role of major evolutionary developments, often in the wake of mass extinctions, that defined the possibilities for future developments of the ecosystem. I have drawn on many sources for the ideas discussed in the following sections, but two papers have proven particularly important.

First, in studying climate shifts I have relied heavily upon Scotese and colleagues' paper summarizing the current state of knowledge concerning Phanerozoic paleotemperatures (Scotese et al. 2021). This provides, among other things, the basis for Fig. 1.3, to which I will refer repeatedly in the text. This shows the oscillation of global average temperatures around the key value of 18 ° C, which defines the switch from hothouse to icehouse conditions. It is particularly useful in the context of the discussion of the Phanerozoic, as it shows the relationship between the link between bifurcations in the curve and chronostratigraphic boundaries. In some cases, these switches occur at major boundaries, such as between the Permian and the Triassic, which is also the boundary between the Palaeozoic and the Mesozoic. In other cases, the shift is associated with lesser boundaries, such as the switch between the Mississippian and the Pennsylvanian, which divides the Carboniferous into two epochs. These changes are far too complex to describe as simple causal relationships. No reasonable argument can be made that the development of an ice-cap creates a sufficiently dramatic change in conditions that a new period or epoch can be defined. Rather, it indicates that any system changes of sufficient magnitude to either trigger the creation or the collapse of permanent ice-caps, are also going to change a lot of other things. It turns out that there are several different likely drivers of a shift from hothouse to icehouse conditions, and vice-versa. Some are internal feedbacks, others are external circumstances. The consistent factor is that it is a big deal when it happens.

In discussing mass extinctions, I have made extensive references to Bond and Grasby's paper on the causes of mass extinctions (Bond and Grasby 2017). There is some ambiguity about what constitutes a mass extinction event, since the idea was first explored in detail by Sepkoski (1989). Further, there is not always general agreement on how severe any given extinction event was. For example, Bond and Grasby note that four different authors variously attribute the End Permian extinction as culling 83%, 62%, 58%, and 57% of marine genera. That's quite a big range. Rather than bog down in this disputed terrain, they have focussed on 21 events for which than can identify a credible "kill mechanism" in the geological record. There are many other events that they dub as biodiversity crises, when anything from 3% through to 60% of genera were lost. The resulting summary graph, Fig. 2.2 gives a useful perspective on the pattern of Phanerozoic extinctions. Bond and Grasby conclude that most mass extinctions are associated with extensive and protracted periods of volcanic activity, though some are associated with bolide impacts – extra-terrestrial objects impacting the Earth. It is worth noting, however, that many events have no currently known cause. Again, I will refer to this figure frequently throughout the chapter. However, as with the climate curve, causation is complex, and correlations in time are often fuzzy (though there are some notable exceptions to this). It is possible to cross-tabulate between some of the major shifts in climate, and some of the extinction events. Some are harder to link in time and space. The Permian Triassic boundary had a major extinction event, a possible bolide impact, and a switch from icehouse to hothouse conditions. The Cambrian extinction events are a bit less clear-cut. Mapped onto figures drawn at a sufficient scale to fit into this book it's possible to "see" possible, and even probably matches between some of the rapid fluctuations in the prevailing hothouse conditions of the time, and the extinction events. However, the hundreds of thousands of years covered by the width of each line muddy the waters a little.

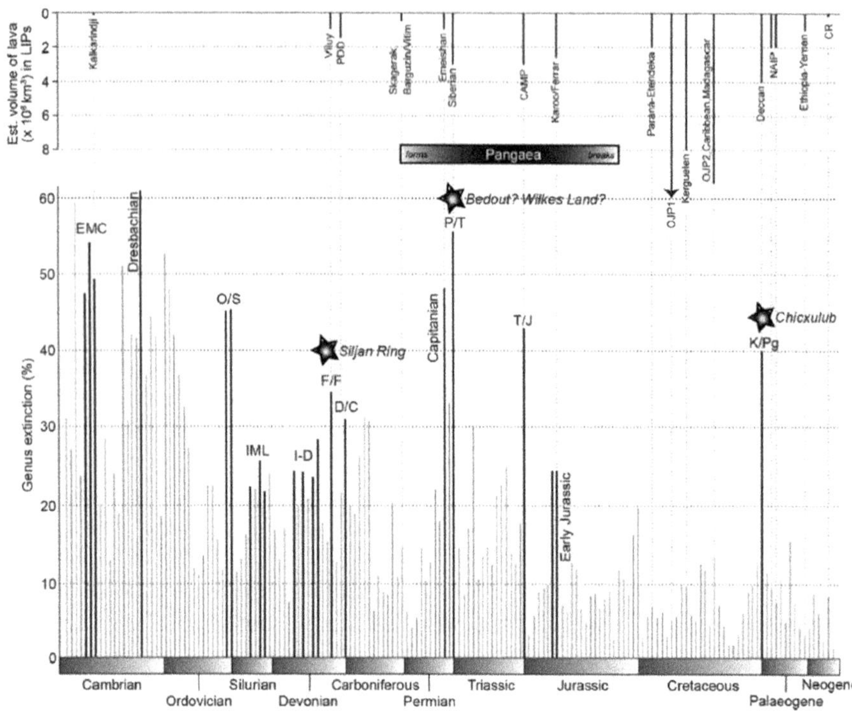

Fig. 2.2 Phanerozoic extinctions plotted against large igneous provinces and known or hypothesized bolide impacts. (From Bond and Grasby 2017)

In discussing the evolutionary advances typically accompanying boundaries in the chronostratigraphic record I have resorted to the tried-and-true mnemonics of the undergraduate curriculum. The Devonian is the age of fishes; the Triassic is the age of Crocodiles; the Paleogene is the age of mammals, and so on. For this I have relied on a variety of sources, including some classic geology textbooks such as Grotzinger and Jordan's *Understanding Earth* (Grotzinger et al. 2010), and some more specialised literature, such as Rose's *The Beginning of the Age of Mammals* (Rose 2006), and Klein's *The Human Career* (Klein 2009).

These broader surveys are useful for the following sections as, although debates and questions remain about innumerable details of the Phanerozoic, these more general sources provide the broad structure needed to navigate the next 541 million years. This noted, the chronologies of this eon are constantly being refined, and palaeontologists have repeatedly demonstrated that boundaries previously considered definitive, are incorrect. For example, although the Devonian is "The Age of Fishes" there *were* fish in the Ordovician. There were mammals in the Cretaceous. There are (avian) dinosaurs in the Cenozoic. And so on. Regardless of our constantly improving knowledge, and these shifts in understanding, the general rule holds: a new chronostratigraphic division, at whatever scale, trims the web of life at many levels. The more taxa that are removed, the greater the opportunities for evolution.

2.4 Paleozoic Periods

The Paleozoic era, running from 541–251.9 Mya, is divided into six geological periods, three of which are distinguished from one another by major extinction events when significant proportions of life on Earth perished as a result of some substantial change in environmental conditions. Whereas in the Precambrian, cataclysmic and transformative boundary events often played out over hundreds of millions of years in the Phanerozoic the pace of the story changed and accelerated. Boundary events dividing periods took much less time to play out. Shifting Earth systems from one state to another could occur within a few hundred thousand years, rather than taking millions of years. A large proportion of Earth's visible geology comes from this time forwards. It is also the time in which key elements of the planet's present form came into being. Such fundamental things as vascular plants, terrestrial vegetation, fishes, and the bilateral symmetry of vertebrates were all established as major evolutionary pathways during this 289.1-million-year span.

2.4.1 Cambrian

The Cambrian, from 541–485.4 Mya saw an extraordinary set of transformations in practically all Earth systems. The rapid increases in available oxygen and mineral nutrients that marked its beginning occurred contemporaneously with shifts in the most fundamental characteristics of individuals and ecosystems. The plants and animals that survived the end of the Proterozoic were joined by much more complex organisms. Bilateral and radial symmetry, shells and exoskeletons, complex digestive systems, and eyes all emerged as successful adaptations. This has led the paleontologist Richard Fortey to describe the one of the Cambrian's defining fossil classes, the trilobites, as the "eyewitnesses to evolution", as they were probably the first creatures to see the world around them (Fortey 2010). This proliferation and diversification of body forms and behaviors went hand-in-hand with the development of increasingly complex predator-prey relationships, as an "arms-race" of adaptations emerged between players in an complex world. The great variety of life was confined to the sea, which at that time was the single ocean Panthalassa, covering most of the Northern Hemisphere.

Identifiable components of contemporary geography that developed during the Cambrian, were the four major structural elements of the North Atlantic region: Laurentia, Baltica, Amazonia, and West Africa, although they were lumped together into radically different formations. In their various configurations they drifted across the surface of the Earth in several different directions, traveling variously from the poles, through the mid-latitudes, to the equator, and then back again. The shifts and rearrangements of geography changed atmospheric and oceanic circulation, and the movements of fresh water and sediments, and so also defined the parameters of the increasingly complex ecologies. At the beginning of the Cambrian, Iapetus, the ocean that separated these continental fragments, started to close. The end of the Cambrian was marked by a significant extinction event, possibly as a consequence of volcanic activity driving significant cooling and a reduction in oceanic oxygen, though no culprit LIP has so far been identified. Regardless of its cause it was of sufficient magnitude to mark the boundary between two periods, and to create significant new ecological space.

2.4.2 Ordovician

The Ordovician, from 488.3–443.8 Mya, opened with a boom of evolutionary activity, the Great Ordovician Biodiversification Event (GOBE), and ended with another major extinction event. The phyla that had evolved previously in the Cambrian were progressively populated with new classes, orders, and families. Life was still almost exclusively confined to the oceans throughout the majority of this period, however within this environment new ecosystems emerged. The most common organisms such as trilobites, corals, sponges, algae, and bivalves became linked into ever more intricate relationships. New survival strategies evolved as things became crowded. This propelled two important jumps: the first vertebrates – jawless armored fish – evolved, and; during the last few million years the first land plants started to colonize the hitherto barren continents.

At the same time Laurentia, Baltica and Avalonia continued to move towards one another, finally closing Iapetus. The collision crumpled ocean floor sediments into a web of mountains. These folds form parts of several contemporary ranges including the Appalachians, the Ozarks, and the Atlas Mountains. When finally assembled the new continent, Euramerica straddled the equator. At its heart was the sutra joining the Americas, Africa, Europe, the Caribbean, Scandinavia, and Greenland. The underlying structure of almost the entire NAPT, and the majority of its topographical armature originated in the mountains and valleys of this collision zone. The only other landmass at this time was Gondwanaland.

The end of the Ordovician is typically identified as the first of the "Big Five" extinctions. Although the causes and kill mechanisms are still uncertain, an extensive volcanic eruption has been suggested as a possibility, as "Large Igneous Provinces" (LIP), are associated with other major extinctions (Bond and Grasby 2017). Global cooling may also have been a driving force. As Gondwana moved over the South Pole it froze, initiating the first of the two major glacial periods in Paleozoic. The ice-covered continent changed the Earth's radiation budget, lowering sea levels as water was bound up in ice caps. An increase in volcanic activity or a bolide impact has also been suggested as a possible cause. Whatever the case, complex Ordovician ecosystems collapsed as warm shallow water habitats disappeared, the amount of photosynthesis declined, and dissolved oxygen in the oceans dropped to levels fatal for many organisms.

2.4.3 Silurian

The Silurian, from 443–417 Mya, opened with a warming trend that drove glaciers into retreat, and initiated a return to higher sea-levels. The ecological response was the familiar burst of evolution, diversification, and consequent increasing complexity. The taxa that survived the end Ordovician extinction event multiplied, divided into new species, and expanded into new habitats. Thus, in the oceans fish evolved jaws, and moved into freshwater environments, while the first corals evolved. Vascular plants, the basic building block of all subsequent terrestrial ecologies started to colonize continents, and their respiration started to affect atmospheric chemistry (Harrison and Morris 2018). This created a positive feedback loop which accelerated the warming trend, and so speeded up the melting of the ice caps. As plants spread their growth medium, the first soils also formed. They also provided the ecosystems into which the first terrestrial arachnids and centipedes were born. The supercontinent Gondwana still covered a large part of the southern hemisphere from the South Pole to the Equator. Two other large land masses existed, Siberia and Laurentia. They were separated from Gondwana by three shallow, warm, island-studded, oceans. The Paleo-Tethys, the Rheic, and the Iapetus. The Panthalassic Ocean covered the rest of the Earth's surface. By the second half of the Silurian there was a return to icehouse conditions, as cooling drove the formation and expansion of ice-caps, a drop in sea levels, and a series of small extinction events. The Silurian seems to have ended not with a bang, but with a sizzle, as global temperatures dramatically increased.

2.4.4 Devonian

The Devonian, from 419.2–358.9 Mya, opened with a retreat of the ice sheets and glaciers, and a transition to hothouse conditions. Three defining features of the period are a proliferation of fish, the greening of the continents, and in parallel with this, the development of soils. These developments in turn paved the way for the development of more complicated terrestrial ecosystems, creating a habitat into which the evolutionary progenitors of the first land animals, lobe-finned fish, could ultimately move. Shifting geography further contributed to landscape change as Gondwana started to break apart. The supercontinent still dominated the southern hemisphere, but now a northern continent of Siberia, and a smaller landmass of Euramerica, formed. The climate was generally warm, with a couple of brief icehouse interludes. Consequently, sea levels were comparatively high. Early Devonian marine ecosystems were quite localized, with many species remaining in one biogeographical region for long periods of time. However, towards the end of the period there was a "Great Interchange" with what one writer describes as a cosmopolitan fauna spanning the globe (McGhee et al. 1997). Over the course of the period, as life expanded in range and complexity on the land and in the sea, Iapetus started to narrow, as Gondwana and Laurasia collided.

The Devonian ended with the second of the "Big Five" extinctions, with about 75% of species lost in several pulses. The largest was the Kellwasser event (372.2 Mya), when over 70% of marine species were extinguished. The broader Hangenberg event (358.9 Mya) impacted both marine and terrestrial species. These were slow moving transformations, rolling forwards over millions of years. Several causes have been proposed. Increasing numbers of land plants significantly changed atmospheric chemistry by reducing the amount of CO_2. This would have induced cooling, increased the size of ice caps, and reduced sea levels. It has also been suggested that another major volcanic event (LIP) in what is now Ukraine may have contributed to a change in atmospheric chemistry, possibly increasing CO_2, which would have induced warming. This, coupled with reduced ocean oxygen, ultimately drove many Devonian ecosystems to catastrophic tipping points.

2.4.5 Carboniferous

The Carboniferous, from 358.9–298.9 Mya, saw an evolutionary expansion most noticeable on land and among amphibians and insects. Hothouse conditions persisted for the first half of the period, and terrestrial ecosystems were dominated by extensive rain forests. After accumulating for 60 million years the resulting biomass, now fossilized, forms a significant proportion of the world's coal. The diverse range of invertebrates and vertebrate amphibians and terrestrial tetrapods took advantage of the historically high levels of atmospheric oxygen by growing to unprecedented size.

The major tectonic event of the Carboniferous was the collision of Gondwana with Laurasia to form Pangaea about 355 Mya. The kidney-shaped supercontinent curved from the Antarctic almost to the North Pole. The vast Panthalassic Ocean, still extended from the North Pole almost to the South Pole. This binary world of either land or water experienced a drop in sea level as around a third of the way into the period as there was a dramatic shift to icehouse conditions.

The period came to an end with the Carboniferous Rainforest Collapse (CRC) around 305 Mya, when the protracted cooling finally wiped out the tropical vegetation. Although this is counted as one of the smaller extinction events it culminated in a transition to much drier conditions over much of the land-surface, with only isolated islands of rainforest surviving. Its cause is still debated, but it closely matches in time the eruption of a LIP in what is now the Skagerrak Strait, between Denmark and Sweden. This extensive area basalt formed within the right timeframe to have helped alter atmospheric chemistry sufficiently to cause a cooling trend that lasted throughout the Permian.

2.4.6 Permian

The Permian, from 298.9–251.9 Mya saw the last fragments of land not already lumped into Pangaea joined onto the supercontinent. As noted, the icehouse condition that were established in the second two-thirds of the Carboniferous continued until the end of the Permian. In the supercontinent's interior, without the temperature regulation of the ocean, climates were extreme, with large seasonal and latitudinal variations, from extreme heat to extreme cold. The ocean, Panthalassa, extended over 180° of latitude, and 90° of longitude at the equator. The world was divided between one land, and one water with an extraordinary diversity of interconnected environments in each one. Tetrapods rapidly diversified in terrestrial environment to include the ancestors of crocodiles (pseudosuchia), dinosaurs, and mammals. New kinds of insects and fungi proliferated, and in at least some places vegetation was lush enough to form new coal measures.

The Permian period, and with it the Paleozoic eon, ended around 251 Mya with the Permian-Triassic (P-T) extinction event, sometimes called "The Great Dying". This was the third of the "Big Five" extinctions. Over a geologically brief period of around 50 ka years more than 90% of all marine species and 70% of terrestrial vertebrate species became extinct. 57% of all biological families and 83% of all genera were lost. This was one of Earth's most dramatic extinction events to date, and it has been suggested that at this time life on Earth came perilously close to being completely extinguished. It was driven by an extensive series of volcanic eruptions. Over the course of 200 ka years a great deal of basalt erupted in the equatorial region of Pangaea, now Siberia. The LIP known as the Siberian Traps comprises millions of cubic kilometers of lava poured over the landscape. Some gasses emitted from the eruption mixed with water to produce acid rain, CO_2 was a potent greenhouse gas. This process was almost certainly accelerated as methane, a potent greenhouse gas, was released from high latitude melting permafrost, and from rapidly heating oceans. In addition to the climate trend, there was a stagnation of ocean circulation, resulting in a loss of mixing, and so a loss of oxygen. This "runaway greenhouse" effect ended the Permian and also the Paleozoic.

2.5 Mesozoic Periods: Triassic, Jurassic, and Cretaceous

The Mesozoic, the eon of "middle life", lasted for 186 million years, from 251.9–65.5 Mya. It opened with barren desert conditions, anoxic stagnant oceans, and a very sparsely populated planet. Over its course it saw the development of a whole array of new ecosystems, culminating with the reign of the dinosaurs. The world map at the end of the Mesozoic was recognizably composed of the major elements of contemporary geography. Despite significant climatic variation over its course, hothouse conditions persisted for the entire time. It ended with the catastrophic K/T (Cretaceous-Paleogene) extinction event.

2.5.1 Triassic

The Triassic lasted from 251.9–201.3 Mya. It was a hothouse period, with the highest global average temperatures of the Phanerozoic, and it ended with the fourth of the "Big Five" extinctions. Despite the harsh environmental conditions at the period's outset, the catastrophic end Permian extinctions had once again opened new ecological spaces. The gaps created by the Great Dying were filled by a proliferation of new species, particularly among the recently evolved crocodilians, dinosaurs, and mammals. Other new animal forms that emerged included aquatic and flying reptiles, snakes, turtles, and frogs.

Geography was dominated by the gradual fragmentation of Pangaea along the same mountainous sutras that formed when it was assembled. The initial rift formed along a line separating what is now New Jersey from Morocco. The resulting upwelling of magma formed a series of LIP basalt intrusions, a fragment of which makes up the Hudson River Valley's Palisades.

This eruption may well have been the driving force in the Triassic-Jurassic extinction event. Around 30% of marine species, 60% of terrestrial plant species, and 40% of terrestrial species were lost at this boundary. The familiar kill mechanism was the combination of sea level change, ocean acidification, climate change, and broad scale ecological collapse.

2.5.2 Jurassic

The Jurassic, from 201.3–145 Mya, is the best imagined of all geological periods, lending its name to the celebrated book and film franchise. Long before Michael Crichton published Jurassic Park, interpretations of Jurassic fauna showed up on everything from kid's pajamas to gas company logos. Fiction notwithstanding, this was when dinosaurs diversified into a wide array of the recently vacated ecological niches on land, in water, and in air. The evolutionary process was accelerated by the increase in habitat diversity as deserts were replaced by rainforests and the rifting continent increased the length and complexity of coastal environments.

The geography of the present-day world emerged in nascent form over the course of the Jurassic. As Pangaea broke apart an ocean formed in the gap between the Americas, Africa, and Europe. The continental structures of Africa, Europe, the Americas, and Greenland, while differently spaced, were arranged roughly as they are now, and they started moving (approximately) along their current trajectories.

The end of the Jurassic was one of the less dramatic period boundaries. One feature was the occurrence of a "cool interval", made up of over 16 distinct cooling and warming events, though Scotese and colleagues conclude it never reached full icehouse conditions (Scotese et al. 2021). While the middle of the Jurassic and middle of the Cretaceous are clearly sufficiently different to be recognizable as separate periods, the hard line separating them has yet to be defined. It is the only period which currently lacks a "Golden Spike". Rogov and colleagues summarized this, noting the: "Jurassic-Cretaceous boundary is the last system boundary within the Phanerozoic which is not fixed by GSSP… Among the most important problems in the delimitation of this important boundary are the … absence of remarkable changes in fossil assemblages and poor corresponding of changes in assemblages belonging to the different fossil groups" (Rogov et al. 2010: 13).

2.5.3 Cretaceous

The Cretaceous ran from 145–65 Mya. During this period Pangaea completed its breakup into recognizably modern continental blocks, which continued moving on the trajectories that brought them into their current configuration. The Jurassic cooling trend continued for a short time, though temperatures rose for much of the rest of the period, keeping the Earth in hothouse conditions throughout. It ended with the fifth (and best imagined) of the "Big Five" extinctions. In terrestrial ecosystems the number of flowering plants increased. Mammals were slowly becoming more common, although the remained bit-players in the ecological drama. The apex predators were still dinosaurs, most spectacularly the celebrated *Tyrannosaurus rex*. the progenitors of modern feathered birds diversified and increased in numbers. Marine ecosystems also diversified, adding numbers of sharks, rays, marine reptiles, and diving birds.

The Cretaceous ended with the K/T extinction event, which started with the Chicxulub impact, when a meteorite struck the northernmost edge of Mexico's Yucatan Peninsula. The initial impact created shockwaves and tsunamis over hundreds of square kilometers. The ejecta, material thrown up by the impact, almost immediately blocked so much solar radiation that there was a catastrophic cooling. This created a brief icehouse terminus to the otherwise unbroken Mesozoic hothouse conditions. Other consequences of the impact included high temperature falling debris, forest fires, and earthquakes. It wiped out about 75% of all species on Earth, with the death toll disproportionately affecting terrestrial environments, where around 90% of terrestrial animals perishing. The list of animal survivors included representatives of mammals, birds, amphibians, reptiles, and turtles. Freshwater and marine species fared better, particularly closer to the base of the food-web. The few animal land species that survived tended to be small-bodied omnivores and scavengers, a group that included a few of the previously marginalized mammals. The end of the Cretaceous was so dramatic that it also marks the end of the Mesozoic era, and the start of the Cenozoic, the era of "new life".

2.6 Cenozoic Periods: Paleogene, Neogene, and Quaternary

The Cenozoic, lasted from 65.5–0.012 Mya. Following a brief and anomalous icehouse interlude resulting from the Chicxulub impact winter the climate quickly returned to hothouse conditions. These persisted for almost half of the Cenozoic, from 65–39.4 Mya. During this time as recognizably contemporary systems of atmosphere, oceans, geography, and biosphere gradually began to form, and mammals became the ascendant class of animals. In the opening paragraph of "The Beginning of the Age of Mammals Kenneth Rose summarizes it like this:

> Mammals are among the most successful animals on earth. They occupy every major habitat, from the equator to the poles, on land, underground, in the trees, in the air, and in both fresh and marine waters. They have invaded diverse locomotor and dietary niches, and range in size from no larger than a bumble bee (the bumblebee bat … 2 g) to the largest animal that have ever evolved (the blue whale … 100,000 kg). Just over a decade ago, the principal references recognized … 4,629 extant mammal species in … 26 orders. The most recent account now recognizes 29 orders of living mammals … with more than 5,400 species in 1,229 genera… But many times, those numbers of genera and species became extinct. Indeed McKenna and Bell (1997) recognized more than 4,000 extinct mammal genera, many of which belonged to remarkable clades that left no living descendants. The great majority of extinct taxa are from the Cenozoic, the last one-third of mammalian history. (Rose 2006: 1)

Over the course of the Cenozoic and across all of these diverse environments some mammalian species rose to the top of the food web with few if any competing apex predators. For this entire time the Atlantic has been opening like the folds of an accordion, along a line from the Fram strait, between Greenland and Norway, and the South Atlantic. At the same time Africa has slowly been pushing northwards into Europe, while the two Americas have been slowly compressing the Caribbean by converging at a comparable pace.

The three periods of the Cenozoic, the Paleogene, the Neogene, and the Quaternary, are, by virtue of their close historical proximity to us, far better known and more comprehensively studied than any previous point in Earth history. There is much more evidence in the landscape, and Earth surface processes worked in ways much more like they do today.

2.6.1 Paleogene

The Paleogene, from 65.5–23.03 Mya is divided into three epochs, the Paleocene, Eocene, and Oligocene. Its geography, although recognizably modern, had some important differences from today's landscape. North and South America were still separated. Persistent hothouse conditions prevented the formation of permanent ice caps at the poles until the end of the Eocene, and so sea level was significantly higher than at present. Consequently, much of the central part of North America was flooded by a shallow ocean. There were significant inundations in South America, and West Africa. India had not joined the rest of Asia; there was a land-bridge across the northern end of the Atlantic joining Europe and Northern North America (Laurentia); Australia was much closer to Antarctica.

In the ecological space opened up by the K/T extinction survivors multiplied and diversified. As forests spread to the poles, open niches on land and in the water were exploited by ascendant mammals. There was a similar rapid diversification of the only surviving dinosaurs, the birds. Although flight had previously developed as a strategy among some dinosaurs, and possibly even among early birds, the anatomy needed for modern bird flight evolved in the Paleogene. Birds also adapted to fill other niches, evolving into flightless predatory running birds, and aquatic diving birds. Rapid evolutionary diversification also occurred among the surviving reptiles, amphibians, insects, and fish.

Continuing warming through the Paleogene accompanied an extinction most pronounced in deep ocean environments. As many as 50% of marine microorganism species lost in an unusual "bottom up" event (Kennett and Stott 1995). The primary cause was a large release of carbon into oceans and the atmosphere, and the accompanying acidification, and warming. The source of the carbon was probably volcanic activity associated with the North Atlantic Igneous Province, parts of which can be found in Iceland, the Faroes, Greenland, Fingal's Cave and the Giant's Causeway, in Northern Ireland (Kender et al. 2021). The resulting ecological space allowed new mammal families to emerge, including ancestral primates, elephants, hoofed animals, and whales.

The Eocene ended with the Eocene–Oligocene extinction event, sometimes known in Europe as the *Grande Coupure*, or Great Break. A significant number of European mammals became extinct, and were replaced by Asian counterparts. An important component of this was the initiation of icehouse conditions, which have persisted, with a couple of brief hothouse interludes, until the present day. No single cause of this has been identified, though some combination of volcanic activity and meteorite strikes has been suggested. The spatial and temporal discontinuity of the changes is sufficient to indicate that a combination of circumstances drove a drop in global temperatures that impacted different regions and ecosystems in different ways, depending variously on their fragility, the nature of the change, and the local timing of the shift.

The gradual drawing together of the "exploded diagram" map of the world continued, with continental tectonic trajectories moving along their current paths. Some of the Earth's youngest and highest mountains, the Himalayas, and the Andes, were formed from this process. Terrestrial and marine flora and fauna were increasingly dominated by species closely related to modern counterparts. Although there were some exceptions, environmental characteristics from atmospheric oxygen levels, through temperature, humidity, and ecosystem distribution were recognizably modern. Mammals comprised an increasing large proportion of terrestrial animals, and made up a significant proportion of marine animals. Of particular importance to our present circumstances: almost all of primate evolution took place from this time forwards, in predominantly icehouse conditions.

Differences in Oligocene geographical connectivity between continents and oceans created opportunities for species migration. India and Australia were isolated. North Africa was connected to Europe via Iberia, and Asia. There was a minor path between North and South America, and there was a viable Arctic route from Europe to North America. This allowed for both alternate migration paths, and different patterns of ocean circulation. Rising sea levels inundated a large part of Europe, turning much of it into an archipelago for a while.

2.6.2 Neogene: Miocene and Pliocene

The Neogene, from 23.03–2.58 Mya, despite having a brief hothouse period, the "Mid Miocene Thermal Maximum", was dominated by cooling. Icehouse conditions were established around 11 Mya, and the resulting ice caps persist to the present day. The combination of continuing tectonic movement, and lowered sea levels closed some ocean gateways. For example, the Arabian Peninsula collided with Eurasia, shutting the passage between the Indian Ocean and the Mediterranean. Around 5.96 Mya the strait of Gibraltar closed, and the Mediterranean evaporated into a few small salt lakes for around 630 years. By 5.33 Mya this process was reversed, the strait reopened, and the waters of the Atlantic re-entered the Mediterranean Basin in a spectacular event now known as the Zanclean Flood.

By the end of the Neogene several important modern climate and ecological systems were in place. The Polar ice-caps were established. The Himalayas had reached a high enough elevation that they provided the thermal engine establishing the monsoon circulation. Recognizably modern plants, and ecosystems were widespread. Two new biomes, kelp forests and grasslands emerged, each supporting unique ecosystems that are essential components of the modern world. The first apes evolved the hominid line started to diverge from other primates. Early in the Pliocene our first known ancestors, the *Australopithecines*, evolved. The exact relationship between these first hominids and modern humans (and indeed other species of hominids) is still debated. Regardless of the details, we can deduce that prevailing conditions were sufficiently similar to the present day that animals with physiological characteristics very like ours could survive and prosper. More broadly stated, species that lived at that time lived in a world similar enough to our own that most have surviving descendants in the present. By the end of the Neogene the essential structure of the relationships between landscapes, ecosystems, and environmental energetics matched their modern counterparts.

The Miocene ended with dramatic cooling. As the ice caps grew into lower latitudes, their increasing reflectivity exacerbated the trend. Icebound high latitudes and the disappearance of surface ecosystems at the poles marked the end of the Pliocene.

2.6.3 Quaternary: Pleistocene

The Quaternary covers the most recent 2.5 million years of Earth history, and is divided into two ages, the Pleistocene and the Holocene. This chapter ends with a discussion of the Pleistocene. It is important for a couple of reasons. First, because it represents the last major chronostratigraphic division prior to the Holocene, consequently it is the one of which we have the best understanding. Secondly, because during the Pleistocene all known hominids evolved, descending from *Australopithecines*, who first appears in the fossil record around 4.2 Mya years ago.

The Genus *Homo* has prospered during the Quaternary. Each species has had markedly different characteristics of size, shape, geographical range, physiognomy and behavior. Yet so far, no terrestrial environmental circumstances have defeated the hominid package of survival strategies combining bipedalism, stereoscopic vision, manipulative dexterity, tool making, intellectual complexity, cooperation, communication, and collective cultural memory. This historically unique package has persisted so far because evolution has proven equal to the task of managing shifts in environment in our original African home, and adaptation has allowed the most recent species to spread outwards into new environments.

In the context of this book the details and debates surrounding the mechanics of long-term human evolution and the distinctions between various ancestral species are unimportant. Richard Klein includes everyone in the genus *Homo* in his foundational text "The Human Career" (Klein 2009). More useful here is the general principles of how other humans survived until the evolution of anatomically modern humans, around 200 ka years ago. This is interesting, as it speaks to the broader question of humans' capacity for survival. In some cases, this has turned on pure luck, as it does for a great many species and genera at some point in history. In all other cases survival has relied upon adaptation or evolution. Humans have proven adept at extending the scope of physical adaptation and evolution by embracing geographical, behavioral and technological changes. While elements of these survival strategies show up in other animals, humans, regardless of species, have refined and extended the "non-physiological survival package" much further than any other genus.

With these broad generalizations in place, it is useful to spend some time in a more detailed discussion of the last 200 ka years of Quaternary history, when anatomically modern humans first migrated out of Africa, ultimately superseding all other humans.

The Pleistocene icehouse had at least eleven major glacial events, and several smaller regional ice advances (Richmond and Fullerton 1986). The two most recent events are known respectively as the Penultimate Glacial Period (PGP) and the Last Glacial Period (LGP). The PGP lasted from ~194 ka to ~135 ka years ago; the LGP lasted from ~115 ka to ~11 ka years ago. The start of the PGP overlaps the current best estimate for the evolution of anatomically modern humans, in coastal Southern Africa, around 200 ka years ago. This suggests that environmental pressures may have contributed to the emergence of the new species. Details of when and where the new species of humans evolved are still debated, as are the possible pattern of their dispersal out of Africa (Rito et al. 2019). The significance of *Homo sapiens* is summarized by Klein like this:

> Before the emergence of modern people, the human form and human behavior evolved together slowly, hand in hand. Afterward, fundamental evolutionary change in body form ceased, while behavioral (cultural) evolution accelerated dramatically. A reasonable explanation is that the modern human form—or more precisely the modern human brain—permitted the full development of culture in the modern sense and that culture then became the primary means by which people responded to natural selective pressures. As an adaptive mechanism, culture is not only far more malleable than the body but cultural innovations can accumulate far more rapidly than genetic ones, and this explains how, in a remarkably short time, the human species has transformed itself from a relatively rare, even insignificant life form to the dominant mammal on the planet. (Klein 2009: 615)

People started to move out of the African epicenter sometime between 75 ka and 50 ka years ago, during the LGP. Although the expanding ice was a long way to the north, the global cooling trend changed the environment significantly. As temperatures fell patterns of precipitation shifted and local and regional ecologies were transformed. Simultaneously there is evidence of changing in human behavior. After about 60 ka years ago human behavior became more sophisticated, with changes in both social organization and mental function suggested by the appearance of novel artifacts, such as decorated objects. This may have been a consequence of an evolutionary change in human neurology, and may be linked to the development of "the modern capacity for rapidly spoke phonemic speech" (Klein 2009: 650).

The Late Pleistocene climate fluctuations drove a periodic expansion of tropical ecosystems, pushing arid deserts, broadleaf, and coniferous forests into higher latitudes in both hemispheres. In both Africa and the Americas at times tropical rainforest expanded over much larger areas, and the present-day deserts were vegetated (Adams 1997; Larrasoaña et al. 2013). Globally there was the predictable inverse relationship between ice and sea-level: seas rose as ice melted, falling as water was once again bound up in ice. Overall the high-stand of about 12 m above present levels occurred between 130 ka and 124 ka, though how this manifested locally is less well-known. Some parts of the Atlantic Rim are much better understood than others: there are fairly good models for Europe, but the Americas and North Africa are less well understood (O'Learyd and McCulloche 2007; Otvos 2015; Hammond et al. 2017).

Where evidence is available in the form of raised ancient beaches, distinctive and datable corals, and marine sediments, definitive measurements can be made. Yet, there are many places for which there is no record from even the recent past. Simple "bathtub" models, which add or subtract a sea-level change estimate to the current coastline, assume a flat ocean surface and static continents. This doesn't work well even over modest distances because the ocean is unevenly distributed across the outside of a spinning, constantly changing, and flexible three-dimensional object. Consequently, there are a lot of bumps, divots, and wrinkles on the surface. In higher latitudes, when ice sheets form during ice ages the crust is pushed downwards by the mass of the glaciers. When these glaciers melt the crust bounces upwards. In some cases, such as the British Isles, the removal of an ice sheet on one end causes an upward motion and falling sea level there. There is a seesaw effect that causes a downward movement at the other end resulting increasing sea levels there (Shennan et al. 2002). Further complicating matters, the volume of the ocean increases as meltwater flows into it during interglacials and as temperature increases the water expands. Tidal flows, ocean currents, the shape of the ocean floor, and the shape of the coastline can all constrict and redistribute the ocean's surface still further.

With these caveats noted we can still approximate the changed geography when the ocean was 12 m above its present level. On the eastern side of the Atlantic the Scandinavian peninsular was an island (Miettinen et al. 2014); a significant proportion of the British Isles and the Low Countries were submerged (De Clercq et al. 2018). However, there are indications that in other places the change was smaller. In Mallorca, in the Mediterranean, the difference may have been only 2.75 m (Polyak et al. 2018). We do not have such good information about the Atlantic coasts of Africa and America. Immediately after the peak of warming between the PGP and the LGP, around 123 ka there was a rapid cooling trend and sea levels fell again (Clark and Mix 2002).

In the LGP the Earth's temperature averaged about 6 °C less than at present, and ice sheets covered the majority of land surfaces above 45° latitude in both hemispheres. There was extensive sea ice covering large areas of high latitude ocean and glaciers extended over the higher slopes and valleys of many mountains, even in the tropics. As the ice sheets expanded tundra, boreal forest, and broadleaf forests all migrated to lower latitudes. Tropical deserts moved to lower latitudes, and the rainforest belt shrank (Anhuf et al. 2006). Sea levels fell and as previously, this change was not a simple linear shift. The redistribution of weight on landmasses, and the resulting downward bend of the crust meant that apparent sea level rises occurred in some places.

Over the course of the LGP anatomically and behaviorally modern humans started to move out of Africa. The story of humans' journey is incomplete, and the models are complicated by the fact that the evolutionary connections between *Homo sapiens* and at least two populations of our closest relatives, Neanderthals and Denisovans, are imperfectly understood. Both species had made it out of Africa in the middle of the Pleistocene, before the LGP, around 130 ka years ago, spreading through southern Europe and Asia. We carry DNA from both, as some of our ancestors successfully conceived and bore children with them (Finlayson 2019). Leaving these ambiguities aside, however, the *Homo sapiens* journey can be mapped in broad strokes (Nielsen et al. 2017). People migrated out of Africa and moved into the Levant and Arabia between 120 ka and 90 ka. By 50 ka they had made their way to Indonesia and Australia; by 45 ka they were in Southern Italy (Benazzi et al. 2011); by 20 ka they were in North Eastern Asia; by 15 ka they had crossed the Bering Straits into the North Western corner of North America (Rito et al. 2019).

The technologies used by these adventurous pioneers that survive into the archaeological record develop and change slowly. So much so that the entire period until the end of the LGP is sometimes described as a single cultural phase, the

Paleolithic, although this is typically subdivided into three smaller segments, the Lower, Middle and Upper. It is useful to remember that 94% of human history, from 195 ka to 11.7 ka is in the Paleolithic. Despite the limited data set describing non-transient artefacts, we can deduce that there must have been developments in other aspects of human culture. Survival techniques, hunting methods, and food preparation must have changed in many times and in many ways to contend with new environments and new resources. By the end of the LGP people inhabited ecosystems from tropical savannah, through rain-forests, deserts, temperate woodlands, boreal forests, tundra, to the maritime arctic. It is highly unlikely that any one individual, or even one group of people lived in more than one or two of these diverse environments at a time, even within a single lifetime. Yet collectively humans acquired and mastered a truly mind-boggling array of knowledge.

It is worth taking a moment to imagine the level of technical skill required to, say, identify a new prey species. Then to track, kill, butcher, and eat it as part of a nutritionally balanced diet. In places where food-webs were very different from tropical environments this would require some ingenuity. While the blade technology seems to have been fairly consistent for long periods during the Paleolithic the array of other technologies and the accompanying knowledge base, the vast majority of which are lost to history, is quite extraordinary. For example, in Arctic environments the only reliable source of vitamin C is marine mammal fat. Learning this, and devising an effective strategy to meet the need required a radical rethinking of established practice. Technologies as diverse as fire making, fuel gathering, woodworking, clothing, cordage, and medicine were all redesigned for each new environment. Developing each required systematic research and development. Maintaining and transmitting new methods in turn required a system of information storage and retrieval, and an educational system. The process of technical reinvention was a requirement for human survival and dispersal. However incremental the process of moving into new environments there were numerous critical technical thresholds that humans crossed in their journey out of the tropics and into the LGP.

The LGP provided a series of "push" factors encouraging migration, as the familiar environments of Africa were transformed by climate change. It also provides a series of "pull" factors as fertile corridors opened up across previously inhospitable landscapes. Lower sea levels opened land bridges between islands and continents, in higher latitudes some ice roads opened, and some paths were blocked. This radical alteration of the "cost surface" of movement across the world may explain how people were able to migrate across a wide variety of landscapes, and move into new environments, but it does not explain why.

There is something of a paradox in the migrations of Paleolithic humans. Some communities seemed content to remain in one place when things worked out for them, often for thousands, or even tens of thousands of years. Others migrated into radically new environments, perhaps because things didn't work out for some reason, or when new opportunities opened up for them. The sophistication and simplicity of at least some of the toolkits used for this entire period suggests that, when they felt the need needed they could travel fast and light. People with few (if any) permanent possessions, and small families could just pick up the kids and walk to a new location when the spirit moved them. New stone tools could be easily and quickly manufactured, and food was periodically abundant. The greatest challenges were almost certainly navigating environmental obstacles, such as deserts, oceans, and high mountains, and figuring out survival strategies in new ecosystems. Yet people met these challenges, and continued to migrate into every habitable environment. The pattern only started to change with the start of the Holocene, around 11.7 ka.

References

Adams JM (1997) Global land environments since the last interglacial. Oak Ridge National Laboratory. http://www.esd.ornl.gov/ern/qen/nerc.html

Anhuf D, Ledru MP, Behling H, Da Cruz FW Jr, Cordeiro RC, Van der Hammen T, Karmann I, Marengo JA, De Oliveira PE, Siffedine PL (2006) A Paleo-environmental change in Amazonian and African rainforest during the LGM. Palaeogeogr Palaeoclimatol Palaeoecol 239(3–4):510–527

Beerling D (2017) The emerald planet: how plants changed Earth's history. Oxford University Press, Oxford

Benazzi S, Douka K, Fornai C, Bauer CC, Kullmer O, Svoboda J, Pap I, Mallegni F, Bayle P, Coquerelle M, Condemi S (2011) Early dispersal of modern humans in Europe and implications for Neanderthal behaviour. Nature 479(7374):525–528

Bond DP, Grasby SE (2017) On the causes of mass extinctions. Palaeogeogr Palaeoclimatol Palaeoecol 478:3–29

Clark PU, Mix AC (2002) Ice sheets and sea level of the last glacial maximum. Quat Sci Rev 21:1–3

De Clercq M, Missiaen T, Wallinga J, Zurita Hurtado O, Versendaal A, Mathys M, De Batist M (2018) A well-preserved Eemian incised-valley fill in the southern North Sea Basin, Belgian continental shelf-coastal plain: implications for northwest European landscape evolution. Earth Surf Process Landf 43(9):1913–1942

Drabon N, Byerly BL, Byerly GR, Wooden JL, Keller CB, Lowe DR (2021) Heterogeneous hadean crust with ambient mantle affinity recorded in detrital zircons of the green sandstone bed. South Afr Proc Natl Acad Sci 118(8)

Finlayson C (2019) The smart Neanderthal: bird catching, cave art, and the cognitive revolution. Oxford University Press

Fortey R (2010) Trilobite: eyewitness to evolution. Vintage

Garcia AK, Schopf JW, Yokobori SI, Akanuma S, Yamagishi A (2017) Reconstructed ancestral enzymes suggest long-term cooling of Earth's photic zone since the Archean. Proc Natl Acad Sci 114(18):4619–4624

Grotzinger J, Jordan TH, Press F (2010) Understanding earth. Macmillan

Hammond AS, Royer DF, Fleagle JG (2017) The Omo-Kibish I pelvis. J Hum Evol 108:199–219

Harrison CJ, Morris JL (2018) The origin and early evolution of vascular plant shoots and leaves. Philos Trans R Soc B 373(1739):20160496

Johnson BW, Wing BA (2020) Limited Archaean continental emergence reflected in an early Archaean 18O-enriched ocean. Nat Geosci 13:243–248

Kamber BS (2015) The evolving nature of terrestrial crust from the hadean, through the Archaean, into the Proterozoic. Precambrian Res 258:48–82

Kender S, Bogus K, Pedersen GK, Dybkjær K, Mather TA, Mariani E, Ridgwell A, Riding JB, Wagner T, Hesselbo SP, Leng MJ (2021) Paleocene/Eocene carbon feedbacks triggered by volcanic activity. Nat Commun 12(1):1–0

Kennett JP, Stott LD (1995) Terminal Paleocene mass extinction in the deep sea: association with global warming. In: Effects of past global change on life, vol 15. National Academy Press, Washington, DC, pp 94–107

Klein RG (2009) The human career: human biological and cultural origins. University of Chicago Press

Knoll A (2021) A brief history of the earth: four billion years in eight chapters. Harper Collins

Larrasoaña JC, Roberts AP, Rohling EJ (2013) Dynamics of green Sahara periods and their role in hominin evolution. PLoS One 8(10):e76514

Mauyama S, Santosh M, Azuma S (2018) Initiation of plate tectonics in the hadean: Eclogitization triggered by the ABEL bombardment. Geosci Front 9(4):1033–1048

McGhee GR, Brett CE, Baird GC (1997) Late Devonian bioevents in the Appalachian Sea: immigration, extinction, and species replacements. In: Paleontological events: stratigraphic, ecological, and evolutionary implications. Columbia University Press, New York, pp 493–508

Miettinen A, Head MJ, Knudsen KL (2014) Eemian sea-level highstand in the eastern Baltic Sea linked to long-duration White Sea connection. Quat Sci Rev 86:158–174

Nance RD, Murphy JB (2019) Supercontinents and the case for Pannotia. Geol Soc Lond, Spec Publ 470(1):65–86

Nielsen R, Akey JM, Jakobsson M, Pritchard JK, Tishkoff S, Willerslev E (2017) Tracing the peopling of the world through genomics. Nature 541(7637):302–310

O'Learyd MJ, McCulloche M (2007) Global sea-level fluctuations during the last Interglaciation (MIS 5e). Quat Sci Rev 26:2090–2112

Otvos EG (2015) The last interglacial stage: definitions and marine highstand, North America and Eurasia. Quat Int 383:158–173

Polyak VJ, Onac BP, Fornós JJ, Hay C, Asmerom Y, Dorale JA, Ginés J, Tuccimei P, Ginés A (2018) A highly resolved record of relative sea level in the western Mediterranean Sea during the last interglacial period. Nat Geosci 11(11):860–864

Richmond GM, Fullerton DS (1986) Summation of quaternary glaciations in The United States of America. Quat Sci Rev 5:183–196

Rito T, Vieira D, Silva M, Conde-Sousa E, Pereira L, Mellars P, Richards MB, Soares P (2019) A dispersal of Homo sapiens from southern to eastern Africa immediately preceded the out-of-Africa migration. Sci Rep 9(1):4728

Robert B, Domeier M, Jakob J (2021) On the origins of the Iapetus Ocean. Earth Sci Rev 221:103791

Rogov MA, Zakharov VA, Nikitenko BL (2010) The Jurassic–cretaceous boundary problem and the myth on J/K boundary extinction. Earth Sci Front 17(Special Issue):13–14

Rose KD (2006) The beginning of the age of mammals. JHU Press

Scotese CR, Song H, Mills BJ, van der Meer DG (2021) Phanerozoic paleotemperatures: the earth's changing climate during the last 540 million years. Earth Sci Rev 11:103503

Sepkoski JJ (1989) Periodicity in extinction and the problem of catastrophism in the history of life. J Geol Soc 146(1):7–19

Shennan I, Peltier WR, Drummond R, Horton B (2002) Global to local scale parameters determining relative sea-level changes and the post-glacial isostatic adjustment of Great Britain. Quat Sci Rev 21(1–3):397–408

Chapter 3
The Greenlandian

The Big Change (11,700 – 8200 BP)

Abstract The first of the Holocene's three ages. The concept of the "ح" ("haa") curve. The 3500 years spanning the transition from the LGM to the 8.2 k Event. The four pivotal changes occurring during the Greenlandian: (1) reduction in ice cover and an accompanying change in sea level, (2) significant changes to North Atlantic geography, (3) widespread extinctions, and (4) widespread cultural transformations.

Keywords LGM · Megafauna extinction · Mesolithic · Neolithic · Doggerland · 8.2 k event

3.1 Introduction

The Holocene is divided into three ages, the Greenlandian, the Northgrippian, and the Meghalayan. Each is defined by a marked shift in global average temperature. The overall trend for the whole Holocene resembles the Arabic letter "ح" (typically transliterated as "haa") flipped a quarter turn right (Fig. 3.1 The generalized trend of Holocene climate – the "ح" curve). The Greenlandian constitutes the rising arc, with a rapid temperature increase over 3500 years, terminating in a brief lurch into cooling at its end, called the 8.2 k event. The Northgrippian is the warmer plateau of the curve, again terminating with a sudden cooling episode, the 4.2 k event. The Meghalayan is characterized by falling temperatures, with the current abrupt upturn in global temperatures defining the final upstroke of the ح (Marcott et al. 2013). This abrupt warming at the end of the Meghalayan is one of the key data points used to identify the Anthropocene. I'll address the question of its significance later. As with all such curves, the global trend gives a very broad impression of what was happening and glosses significant variation in time and in space. Marcott's work brings together results from multiple reconstructions that individually give a more nuanced picture of changes. However, here it is the broad pattern that is significant.

M. Bampton, *The North Atlantic Polar Triangle*, Springer Polar Sciences,
https://doi.org/10.1007/978-3-031-27264-6_3

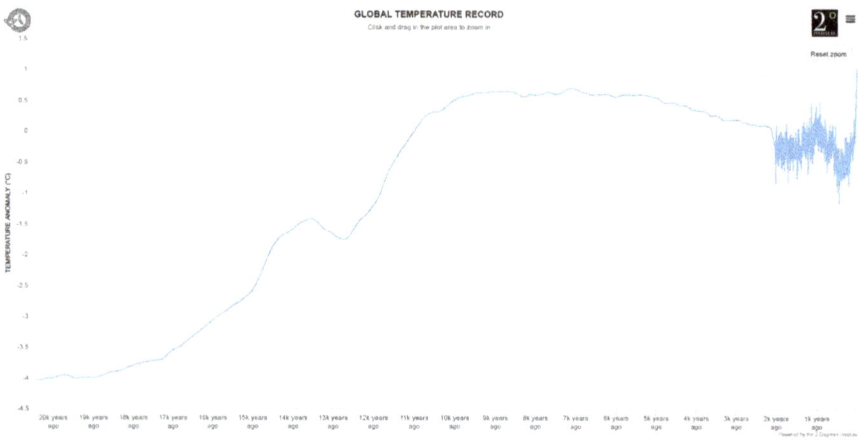

Fig. 3.1 Global average temperature at the end of the Pleistocene, and into the Greenlandian (After https://www.temperaturerecord.org/)

The start of the Greenlandian is defined by the same marker as the start of the Holocene in the NGRIP2 Greenland ice core (Walker et al. 2019). The climate changes within the Greenlandian precipitated changes in numerous other systems. In ecosystems vegetation belts expanded into higher latitudes and altitudes as conditions became milder, and melting ice opened large areas of land to colonization. Some animal populations relocated with them, some collapsed, and others were driven to extinction. Terrestrial megafauna proved particularly vulnerable. The multitude of changes occurring in the environment over the course of the Holocene were intimately intertwined with human social changes. The Greenlandian coincided with the beginning of most profound and protracted period of cultural change in our species' 200,000-year history.

Despite the wide array of environmental and social changes some things were consistent. As we established in the final sections of the last chapter, on this timescale tectonics are essentially stable. That is, while the effects of tectonic processes are visible all around us in the landscape, and there are still events such as volcanoes and earthquakes that are triggered by ongoing tectonic activity, no new supercontinents are formed, no oceans vanish, and no new mountain ranges rise out of the plains. By the time the tectonic processes currently underway have a visible effect on other large Earth systems like climate and ocean circulation, *Homo sapiens* will have been superseded one way or another. Mammal species have an average life span of around one million years from evolutionary inception to extinction (Lövei 2007). It took the Atlantic over 200 million years to open. Even if we do really well and survive for the remaining 800,000 years we have left of our evolutionary life-expectancy we will not see major alterations in global geography resulting from tectonic rearrangements.

What I shall outline here are the trajectories of the four sets of surface processes that underwent significant transformations during the Greenlandian: the coupled climate and ocean system, geography ecology, and society. I shall occasionally stray beyond the NAPT to describe patterns that reach beyond my admittedly arbitrary boundary, as some processes extend out into other areas. However, as much as possible my geographical focus centers on the Atlantic, from the equator to the pole, and is bounded by the Americas, North Africa, and Europe.

3.2 Climate and Ocean

The Greenlandian, the shortest age of the Holocene, lasted only 3500 years. Around 11.7 Kya over a very brief period, perhaps less than a century, temperatures began to rose rapidly. There are good indications that the warming trend started earlier, probably sometime around 19 Kya, but dramatic systemic change occurred only when a critical, and currently unknown, threshold was passed.

The process was driven by astronomical forcing, but once Earth-Sun geometry had initiated the change, a cascade of other physical effects served to exacerbate the process. These included an increase in atmospheric CO_2, increased variability in two major ocean circulation dynamics, the Atlantic Meridional Overturning Current (AMOC), and the Sub-Polar Gyre (SPG). The AMOC is the northward surface transport of warm tropical ocean water, which as it cools in high latitudes is returned southwards at depth. The strength and location of this flow has a direct impact on the amount of energy delivered to the North Atlantic (Hays et al. 1976; Koul et al. 2020). Figure 3.2 shows the latitudinal variation in this warming and indicates its timing. Shakun and colleagues show that once the warming threshold was reached, the AMOC started to see-saw rapidly between increased and decreased strength.

One notable characteristic of this time was a different distribution of insolation – the amount of incoming solar radiation – by latitude. The pattern is clearest in the tropics, between 30° N and 30° S, where radiation was about 2 W/m^{-2} (watts per square meter) lower than at present. In contrast, in the Arctic and Antarctic circles, from 60° upwards insolation was about 6 W/m^{-2} higher than at present. Both of these regions had a distinctive annual pattern. The tropical deficit was most marked during the winter months. In December average insolation 30° each side of the equator was more than 30 W/m^{-2} lower than at present. In June the Northern hemisphere, for which there is the best record, had an increase in solar radiation of between 28 and 36 W/m^{-2} above preset values (Shakun et al. 2012).

The spatial variation in solar radiation translated into a similar variation in temperatures. In South America and Africa, the temperature increase was slower, fluctuated more, and peaked earlier. The highest value was around 0.4° above the series mean by 9000 BP. In the tropics, between 30° N and 30° S temperature curves calculated from proxies in Borneo and Indonesia show a steady rise throughout the Greenlandian, leveling off at 0.4° above the series mean around 6000 BP, well into the Northgrippian. In the Northern hemisphere, from 30° N to the North Pole, temperatures rose sharply, leveling off about 1.2° above the series mean at around 8000 BP.

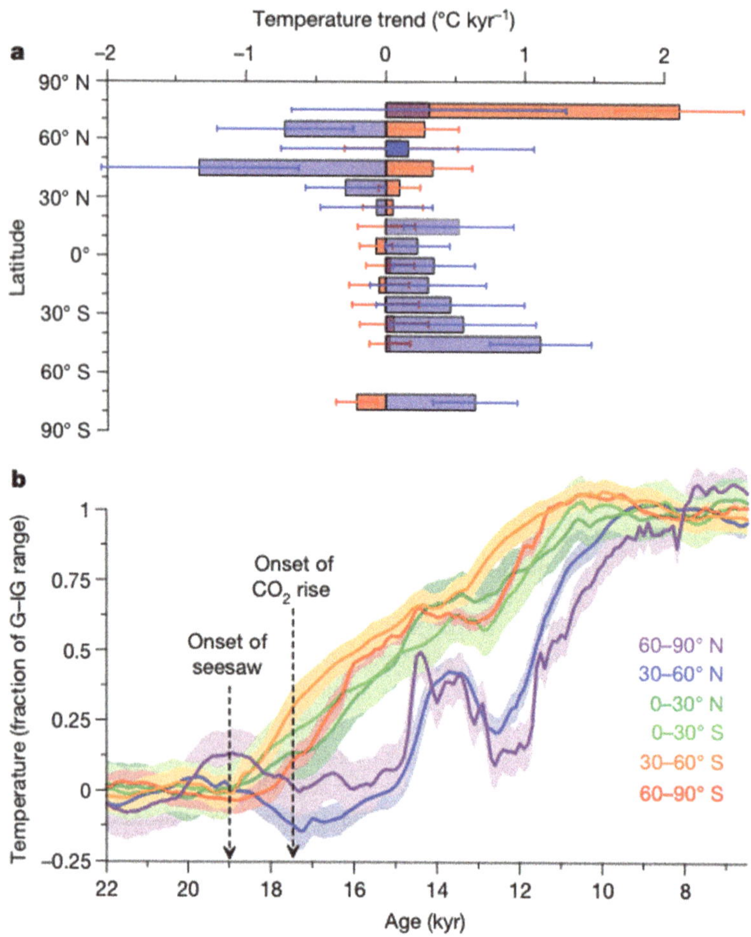

Fig. 3.2 The Onset of Holocene warming (Shakun et al. 2012)

Despite regional and temporal variations, once it was initiated, Greenlandian warming continued steadily and overall global temperatures rose by about 1 °C. At the end of this time the trend was briefly reversed as the last fragments of the Laurentide ice sheet disintegrated. At the same time at least two glacial lakes, Agassiz and Ojibway, released floods of cold water. Combined these events had a potent impact on ocean temperature and circulation. From 11.7 Kya forward the Sub Polar Gyre had extended westwards, interacting with the shrinking Laurentide Ice Sheet. Despite the prevailing warming, until 8.2 Kya there was enough ice remaining to influence climate by slowing the winter westerlies and the sub-polar easterlies, particularly off the coast of Greenland (Gregoire et al. 2018). However, when the last ice collapsed, the meltwater and the outflow from the glacial lakes immediately changed the North Atlantic's temperature and salinity. This, in turn, disrupted the thermohaline circulation altering patterns such as the North Atlantic

Current and the Irminger Current, which typically serve to bring warm water from the tropics towards the Arctic, and then redistribute it between Iceland and Greenland (Colin et al. 2010). The net result was that the prevailing Greenlandian warming trend created the conditions for a short but intense period of cooling, the 8.2 k event.

The Greenlandian ocean had several other major important dynamics. There was an inversion of what we now regard as the "typical" thermal structure of the Atlantic. Currently the ocean is usually warmest at the surface and in the tropics. Temperature typically to decline with increasing depth and at higher latitudes (Cléroux et al. 2013). From the Last Glacial Maximum (LGM) into the Greenlandian there was a millennia-long period during which warmer water at depth was overlain with colder water (Thiagarajan et al. 2014). There is some uncertainty about what caused this. It could have been the result of cold meltwater running off ice sheets. This cold water, buoyant because of its lower salinity, floated over warmer and saltier water flowing northwards from the tropics. As parcels of water with different densities interacted convective circulation produced narrow plumes of water rising from the warmer depths and puncturing the cold surface layer ("thermobaric caballing"). At high latitudes this process causes dramatic warming of the ocean surface and the atmosphere. The net result was melting ice created a pathway for heat energy to first be transported into high latitudes at depth, and then rapidly delivered upwards to the surface. It has been suggested that this may have sometimes happened at speeds of 10 m a minute. This would be sufficient to warming sea surface across the entire ocean basin by around 28 °C within a single month (Su et al. 2016).

The Greenlandian AMOC was stronger than at present, and remained so well into the Northgrippian, finally weakening between 7000 and 6000 Kya (Ayache et al. 2018). It was significantly albeit briefly disrupted by changes in incoming solar radiation, and the influx of cold fresh water during the 8.2 k event. Typically, a weakened AMOC is associated with cooling in high latitudes. This is intuitive – less heat energy heading north results in cooler high latitude oceans. However, present-day weakening the AMOC resulting from melting of the Greenland Ice Sheet is associated with warming (Chen and Tung 2018). This contradiction may simply be a result of scaling effects – a shorter term decadal impact may be overshadowed in the longer-term data by a bigger and more lasting millennial trend. This ambivalence in the model serves as a useful illustration of the difficulties of making historical analogies in environmental modeling. In the Greenlandian the cold freshwater input to ocean circulation from the Laurentide Ice Sheet entered the system at a different place, through the Hudson Strait, into the southern end of the Davis Strait (Fig. 3.3), at about 60°N.

3.3 Geography

The Greenlandian was the last time in which the geography of the North Atlantic differed substantially from its current configuration. This was almost entirely attributable to sea-level changes. The ocean was significantly lower in the midlatitudes,

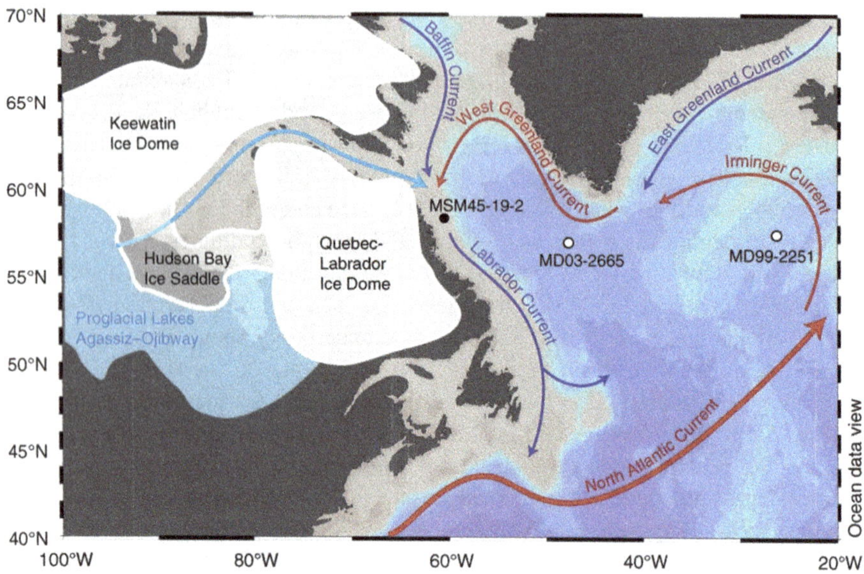

Fig. 3.3 the Collapse of the Hudson Ice Saddle and resulting meltwater discharge through Hudson Strait (After Lochte et al. 2019)

and significantly higher in the Arctic and the tropics. This complex response resulted from a combination of isostatic rebound after the removal of much of the ice burden, the increased volume of the ocean as its temperature rose, and the addition of meltwater from the LGM ice sheet. These effects were unevenly distributed in space and time, with some occurring rapidly, whereas others are still underway today. Crustal rebound occurred in several phases. An initial rapid uplift happened as soon as the ice was removed, more or less keeping pace with melting. This was followed by a much more gradual stabilization, measured in thousands of years. Meanwhile the change in ocean volume occurred fairly rapidly and can be measured in decades and centuries. The major factors influencing ocean volume are, the total mass of liquid water, its temperature, and its distribution. As previously noted, the ocean is not a limpid pool that goes up or down depending on these changes. Rather, it is a wrinkled and poorly fitting mobile wrapping only partly covering the complex and asymmetrical surface of the Earth. The complex response illustrated in Fig. 3.4 shows how this played out throughout the region.

Greenlandian sea-level changes in the North Atlantic can be roughly divided between a rapid drop of >50 m in high latitudes, and a substantial rise of up to 50 m in lower latitudes. These occurred on both sides of the Atlantic, though there are some significant variations that are worth noting. In the west in high latitudes, around 70° N sea levels rapidly fell by over 100 m at a rate of about 23 m/ka. Further south, around 45° N the sea fell by a comparable amount, but significantly earlier, as the ice burden was removed sooner. Between 16 Kya and 12 Kya BP sea level dropped by 125 m at around 30 m/ka. From about 12 Kya forwards this trend

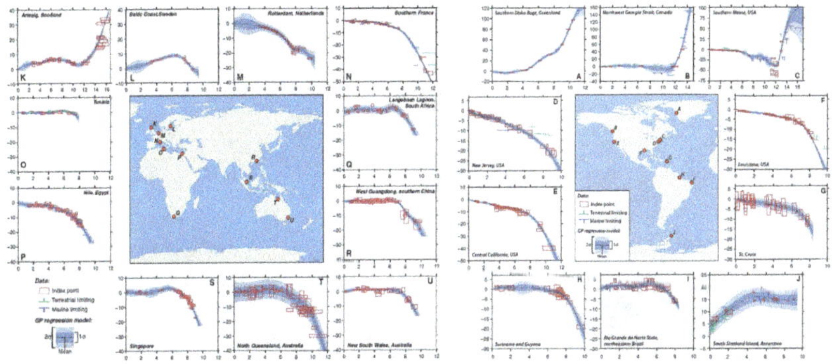

Fig. 3.4 Variations in Holocene sea level change in the Atlantic (Kahn et al. 2015)

had started to reverse, and sea level was rising by 9.6 m/Ka. This slowed to 1.4 m/Ka between 10 Kya and 8 Kya. Further south, around 40° N Sea level rise between 10 Kya and 8 Kya is at about 7.2 m/Ka. On the eastern side of the Atlantic in Scotland, at around 56° N between 12 Kya and 10 Kya sea level dropped by 0.1 m/Ka. Between 10 Kya and 8 Kya this trend had reversed, and sea level was rising by 1.3 m/Ka. In the Netherlands, at 51° N during the same time period the rise was 2.6 m/Ka. At 43° N the rise was 13 m/Ka between 16 Kya and 12 Kya, slowing to 8.6 m/Ka between 10 Kya and 8 Kya. At 30° N, in the Nile Delta, sea level rose at 9.6 m/Ka between 10 Kya and 8 Kya (Khan et al. 2015).

The difference in responses between the higher and lower latitudes resulted from the timing of the melting. Between 40° and 60° N from 16 Kya forwards the crust rebounded as ice started to thin and retreat. Consequently, the sea level dropped rapidly, but by 10 ka sea levels started to rise again, as meltwater caught up with the rising land. Above the Arctic Circle, from 60° northwards, the sea level decline was delayed until around 12 Kya, as it was only then that the ice retreat reached these higher latitudes. South of 40° the sea level rose throughout the Greenlandian, starting around 12 Kya, gradually slowing through time as there was no unloading effect of the ice, only an increase in the amount of available water.

The geographical impact of the changing sea level was particularly marked in the mid and higher latitudes, and the resulting map of the North Atlantic was significantly different from the current one (Fig. 3.5). Scandinavia was united across the mouth of the Baltic, though most of it this bridge was ice-covered until the end of the Greenlandian. There were contiguous ice sheets with centers of dispersal over Scandinavia, Greenland, and Canada in places extending almost as far south as 39°. The Davis Strait remained covered with ice, and was still connected to North America. There were several large islands off the coast of Newfoundland, which remained largely ice-covered. The continental glaciers were connected by sea ice that covered the whole Arctic Ocean, blocked the Fram Strait, sealed the Baltic, and covered the Norwegian Sea. There were also surviving mountain glaciers covering the higher altitudes of interior Britain and Iceland, the Alps, and Pyrenees. At this

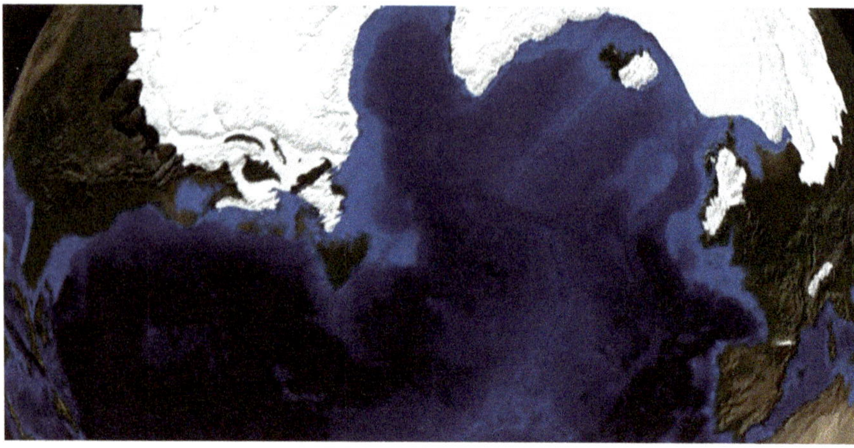

Fig. 3.5 Early Greenlandian Geography in the high latitudes

time an enterprising polar bear could have walked from one side of the NAPT to the other, from Labrador to Denmark hunting along the ice margin the whole way.

South of the ice, in the mid latitudes on the eastern side of the Atlantic what is now the British Isles formed a single contiguous peninsular of Northern Europe. The intervening territory, known as Doggerland, was home to a thriving lowland ecosystem, supporting a Mesolithic European culture. To the west the coast of the USA extended about 100 km further out to sea, reaching the edge of the continental shelf. The outer edge of the Gulf of Maine was defined by the Georges Bank, which was dry land. Unlike Doggerland this now-submerged lowland plain is currently unnamed, and not well explored. However, occasionally artifacts retrieved by modern trawl nets indicate that it had a human population.

In lower latitudes the Florida panhandle was at least twice its present width, extending over 150 km westwards into the Gulf of Mexico. The islands of the Caribbean were also at least twice as large, with many of the outlying keys connected together, to form a much denser and connected archipelago. Nicaragua and Honduras extended over 100 km eastwards. The Amazon issued through a delta extending over 300 km seawards. Although the coast of Africa was closer to its current state, but there was also a seaward extension, particularly along the coast between Sierra Leone and Senegal.

These geographical differences exerted a significant influence on ocean circulation. In addition to the climate-ocean coupling noted in the previous section the ocean was also differently constrained by landmasses now submerged, blocked by ice-sheets now melted, and with a different pattern of bathymetry. In sum, the different geography of the Greenlandian created a significantly different set of ecological and cultural pathways, networks, and barriers. For the first 3.5 Ka of the Holocene the world was differently arranged, with a dynamic and rapidly changing physical environment.

3.4 Ecology

In broad terms Greenlandian ecosystem distribution was determined by the same factors that determine current distributions, though for the last time in Earth history, with much less intentional human manipulation. The ecology of this time is as dynamic as other aspects of the age, manifesting itself in three different areas. First, as ice sheets melted large areas of bare land were exposed resulting in a sequence of ecosystem expansions, migrations, and colonizations. Secondly, oceans that were permanently or seasonally ice-covered transitioned to open water, triggering a transition to new open water ecologies. Thirdly, there was a global wave of extinctions, particularly among terrestrial megafauna, initiating a wave of niche-shifting among surviving species. All three areas were linked, but they unfolded in different ways, and in different time-frames.

The first area of ecological change, the exposure of large areas of new land, had two different trajectories. In one previously ice-covered terrestrial landscapes emerged as almost completely barren when the ice melted. Devoid of soil, frequently scraped down to bare polished rock, these environments developed ecosystems fairly slowly. It takes quite a while to establish a viable biome on a surface devoid of organic matter. The current-day challenges of farming in previously glaciated landscapes such as Northern New England is a legacy of the wholesale removal of topsoil in the LGM. Within the NAPT this includes any land above 40° N in the west, anything above 50° N in the east, and more southerly high mountain regions including the Pyrenees and the Alps (Jaunsproge 2013). The bare land was gradually colonized by opportunistic tundra species, emerging from lower latitude refugia, in a complex multi-step process. Pielou describes reforestation on the East Coast of North America, a process measured in centuries or even millennia, like this:

> Many species of trees invaded from both directions. The populations that followed the mainland route have left evidence of their time of arrival in the form of pollen preserved in lake sediments. They advanced northeastward through New England from their refugia in the south. First to come were poplars, followed by spruces, balsam fir, and paper birch, trees that can endure extreme cold. Then came less hardy trees: oak, maple, eastern white pine, and eastern hemlock established themselves in that order. (Pielou 1991: 215)

In the second ecological trajectory there was a northwards migration of low-latitude biogeographical zones or "belts" that had escaped the expanding ice. These started with the tundra immediately adjacent to the ice-fronts through humid-temperate forests, deserts, and extended to the equatorial tropical rain-forests. This sequence, analogous to the present-day biogeographical zonation, advanced as the ice retreated. A comparable process is now underway as the USDA's "hardiness zones" in North America are migrating northwards. However, the ecosystem transformation of the Greenlandian was more complex than a simple northward shift of everything by 30° or 40° of latitude, and a few hundred meters up the mountainsides. Changes in prevailing climate conditions radically restructured many ecosystems. Differences in humidity and temperature favored some species over others, changing the proportional relationships between plants within ecosystems (Williams et al. 2004).

Those regions that were never ice-covered were reservoirs of biodiversity from which colonizing plants and animals spread as the climate warmed. The refugia served as seeding centers for the expansion of overall biomass. But not everything had survived the cold period, and there was a winnowing of species including a reduction in plant diversity, and a restructuring of relationships between surviving taxons (Zobel et al. 2018). This complex response is summarized in Williams et al.:

> each taxon behaved individually, not as a member of an integrated community. Each appears to have had a different range during the glacial period and to have expanded at a different rate since that time. In line with this, the composition of the vegetation has been changing continuously and there is little reason to believe that any kind of stability has been achieved with the present vegetation pattern. (Williams et al. 1993: 219)

The second area of ecological change, in ocean ecosystems, is harder to outline as knowledge of Greenlandian marine ecology is patchy at best. This is in part because it was, in geological terms, so recent. As there has been insufficient tectonic activity to drain oceans, and uplift ocean sediments, deep-water macro fauna are not yet part of the terrestrial fossil record, as they are for earlier periods. There are no known marine equivalents of the La Brea Tar Pits, or the numerous bog environments, clay deposits, glaciers or high-mountain caves that have yielded preserved non-fossilized and soft-tissue specimens of marine fauna similar to the various Holocene human and animal remains discovered on land. Much of the available literature on prehistoric Holocene marine systems centers on the use of marine proxies to model climate conditions. Core samples of deep-water sediments have yielded a record based on microorganisms, which describe the characteristics of the base of the food web. Higher-level consumers, and full-fledged ecosystem reconstructions remain the stuff of extrapolation, supposition, and speculation. Where ecosystem reconstructions are available they tend to be localized, rather than regional.

Despite this gapped record those studies that are available offer an intriguing glimpse into how these functioned, and provide some general principles can be applied to the less well-studied regions. As with terrestrial systems there were considerable differences in latitudinal responses. The Greenlandian saw rapid and dramatic changes in higher latitudes. The replacement of reflective ice by open water changed the radiation budget by increasing the absorption of incoming sunlight. This effect was somewhat moderated by the cooling effect of fresh melt-water. However, the Arctic was characterized by the coldest and warmest extremes of the Holocene: during the Holocene Thermal Maximum (HTM) peak warmth, around 8 Kya, summer temperatures were 7 °C warmer than today as more solar radiation and warm water reached the Arctic (Van der Bilt et al. 2019). The ice was reduced to such an extent that there may possibly have been some ice-free summers (O'Regan et al. 2011). Further indication that this was a turbulent time for the Arctic, is that these anomalies were not uniformly distributed across the entire region, on the western side North America the HTM happened between 11 Kya and 9 Kya. This was 4 Kya prior to the HTM in northeast Canada. This asymmetry, as previously noted, was attributable to lingering areas of the Laurentide ice sheet, which produced asymmetrical climate patterns throughout the continent (Kaufman et al. 2004). In the earliest part of the Greenlandian, when much of the high-latitude Northeast was still under the ice, there was also extensive sea ice cover and very low local productivity. Between ~9.4 Kya and 8.5 Kya as the sea ice gradually receded and the rate of melt on the Greenland Ice Sheet creased the West Greenland Current became stronger, and there was a matching increase in biological productivity. (WGC) (Saini et al. 2020).

It is reasonable to hypothesize that in these conditions the microorganic building blocks of sub-ice ecosystems extended much further outwards from the continents, matching the greater ice extent. This would have supported an array of smaller predators. Matching sub-Arctic ecosystems would also have been located in lower latitudes. Extrapolating from this to the behavior of wider ranging macro fauna, higher in the food pyramid is difficult, and their behaviors are less well understood. However, research into whale ecology offers some insight into how early marine ecosystems may have functioned. Large vertebrates such as apex predator fish and whales constitute a fairly small proportion of the total ocean biomass. However, to survive the require a large range, and over the long term their bodies, living and dead, constitute significant carbon sequestration (Pershing et al. 2010). Over the course of the entire 11.7 Kya of the Holocene this constitutes a significant component of the marine carbon cycle. The longevity of this component of the carbon cycle, and its continuing significance is evident in whale fall ecosystems.

The larger cetaceans, such as the baleen whales and large carnivorous predators such as sperm whales most frequently die of malnutrition or old age often during migrations between calving and feeding grounds. They have a huge migratory range, often traversing entire hemispheres. The corpses sink to the abyssal plain, often at depths of over 1000 m. These whale-fall sites have unique ecosystems, frequently related to the non-photosynthetic ecosystems of mid oceanic ridges (sometimes called chemolithosynthic systems). The nutritious fat-rich carcasses pass through several stages. First soft tissues are stripped away by scavenger fish and invertebrates. Secondly oil-rich bones are leached of accessible nutrients by micro-organisms. Thirdly, specialized bacteria metabolize remaining nutrients such as Sulphur released by the first two scavengers. Finally, the remaining skeletons structures serve as reefs. This last is of particular significance as some have been speculatively dated to over 10 Kya (Smith et al. 2015). The complex ecosystems associated with the fallen whales is taxonomically linked to the high temperature chemosynthetic ecosystems of volcanically active mid-oceanic ridges and their associated lower temperature "cold-seeps". Similar taxonomic speciation has been found between the Pacific and Atlantic basins (Sumida et al. 2016). These two observations suggest that these ecosystems have ancient origins and are evolutionarily linked. Fossil evidence from at least the Eocene (56 to 33.9 mya) suggests that these systems have a very long history (Shapiro and Spangler 2009).

Other evidence of the early Holocene dynamism of North Atlantic marine ecosystems is visible in the corals of the New England Sea Mounts. Deep sea ecosystems, despite being apparently isolated from surface events are unexpectedly sensitive to seasonal, interannual, and millennial-scale changes. During the LGM North Atlantic corals were confined to a band between around 2000 m and 1700 m. As the conditions warmed and ice receded the corals grew larger and extended their range. (Thiagarajan et al. 2013, 2014).

Another Greenlandian ocean ecosystem change that can be inferred from present evidence is available in the evolutionary history of landlocked salmon. These are salmon that live out their entire lifecycle in freshwater, being born and breeding in streams, and living out the balance of their adult lives in lakes. Their more common sea-dwelling cousins are born in freshwater streams, grow to adulthood in the ocean, returning to their natal freshwater streams to breed. The landlocked salmon were trapped in the freshwater stage of their lives sometime after 14 Kya, when rapidly melting ice and falling sea-levels cut off some drainage systems from the ocean. The landlocked fish are genetically similar to the anadromous fish. When the young are released into river systems that discharge into the ocean, they can adapt to the physiologically difficult transition from fresh- to salt-water (McCormick et al. 2019).

The third area of Greenlandian ecological dynamism, widespread extinctions, has been a contentious issue. There is no questioning the bleak observation made by Araujo "Starting around 50,000 years ago, most large terrestrial animals went extinct in most continents" (Araujo et al. 2017, p. 216). There is abundant evidence to indicate that this phase of terrestrial extinctions continued well into the Greenlandian, as the last populations of such LGM species as mammoths and mastadons died out. Historically the cause has been debated with hyperlethal disease, climate stress, and overhunting all suggested as possible kill mechanisms. The first hypothesis, that in some places, such as North America, a cocktail of hyperlethal diseases, possibly carried from Eurasia by migrating people, may have driven extinctions, has now been abandoned by most researchers. There is, so far, no definitive physical evidence of a disease, and this absence, combined with a shortage of convincing candidates undermines the viability of the idea (Lyons et al. 2004).

The second hypothesis is that climate driven habitat changes pushed species to extinction (Bond and Grasby 2017). There is some evidence supporting this idea. The "plaids and stripes" theory is that there were two fundamentally different adaptive survival strategies demanded by LGM and Holocene conditions. During the LGM habitats changed frequently and radically as climate fluctuated on a decadal to millennial cycles. Rapid and uneven latitudinal shifts in ecological belts created unstable, fragmented, and patchy ecosystems, or "plaids". These caused significant time-lags in adaptation for all resident animals, but the larger creatures enjoyed "greater mobility, lower cost of locomotion, greater dietary breadth, and higher metabolic efficiency" (Mann et al. 2019, p. 334). This gave them a significant advantage in a world where migration in pursuit of favorable conditions was a constant need. The Holocene, in contrast, enjoyed a more complacent climate in which, once established, latitudinal and altitudinal distribution of climate conditions remained stable over many millennia. These were the "stripes". In these conditions big animals' "relatively low population densities… small population sizes … slow rates of reproduction, and the need for large daily food intakes" become liabilities (Mann et al. 2019 p. 334) (Fig. 3.6).

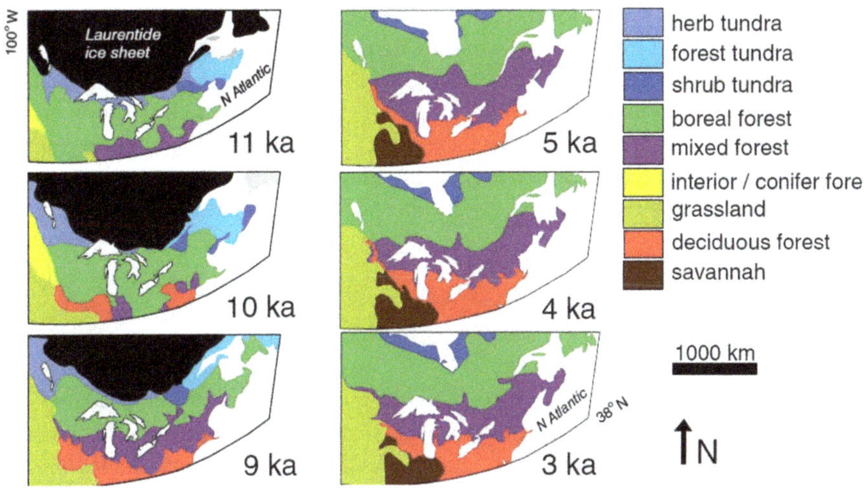

Fig. 3.6 Ancient examples of plaid (left) and striped (right) landscapes (Mann et al. 2019)

The third hypothesis is that interactions with humans triggered the wave of early Holocene extinctions, particularly outside the African continent. There is abundant physical evidence of mass-killing of megafauna in the Americas following its human colonization. This type of hunting strategy had a devastating effect on animals with slow reproductive rates. However, not all megafauna on all continents were wiped out in one huge killing spree whenever humans first showed up. Habitat competition, habitat change, and long-term predation all also contributed to human erosion of large animal diversity. Different extinction times and rates on different continents suggest that there were a number of ways in which humans pushed megafauna to extinction. In Africa a relatively low megafaunal extinction rate of <18% suggests that coevolution with hominids best prepared these animals to cope with all kinds of humans, from Australopithecines onward. In Europe a higher extinction rate of around 29% spread over 40,000 years indicates that long association with both *Neanderthals* and *sapiens* gave some large animals a fighting chance of adaptation to humans. In comparison the Americas experienced between 72% – 88% megafaunal extinction rates closely following humans' arrival on the scene. Comparable rates of extinction in Australia and New Zealand suggest that the arrival of humans is almost certainly a trigger for megafaunal extinction (Araujo et al. 2017). Populations already stressed by rapid changes in environment, as happened at the end of the LGM were particularly vulnerable.

The impact of extinctions on Greenlandian ecosystems was wide-ranging. Megafauna sit at the top of the trophic pyramid. Whether carnivores, herbivores or scavengers, they all act as ecosystem engineers. Their physical bulk demands a big diet and so they transform the world around them in lasting ways over long periods of time (Barnosky et al. 2016). They also support and influence a host of associated species of plants and animals including, variously, parasites, predators, prey, mutualistic partners, and assorted opportunists (Galetti et al. 2018). Each large animal species removed from an ecosystem fundamentally transforms the way it functions, changing trophic pathways and reorganizing evolutionary pressures and possibilities. This process has been described as "trophic downgrading" (Estes et al. 2011). The loss of large-bodied animals reaches far beyond their habitat and range, impacting everything from disease dynamics, fire ecology, carbon sequestration, invasive species, to biogeochemical cycles in soil, water, and atmospheric systems.

3.5 Culture

The Greenlandian saw a transformation of human culture that was unprecedented in its speed, extent, and variety. The two biggest changes of which we have clear record are first, a renewed wave of migration out of Africa as population increased, and secondly the expansion of humans into new parts of Europe, Asia, Africa, and the Americas. Retreating ice sheets, fertile land bridges, and changing ecologies provided opportunities and pressures for migration. Humans, along with many other species expanded into newly survivable and accessible landscapes. As communities were separated from one another by ever greater distances, and moved into new environments, cultures diversified. The archaeological practice of using materials to define cultural grouping, stages, and developments gives an indication of how rapidly this proliferation of social forms occurred at the end of the LGM.

Greenlandian cultures are grouped in Europe and Africa under the blanket term Mesolithic, spanning the time from around 20 Kya to 5 Kya. However, within this broad categorization there are a wide variety of identifiably different lifeways evident in the record. A partial list includes ranges from the Iron Gates Culture of the Balkans (13 Kya – 5 Kya), through the Azilian (14 Kya – 10 Kya) and Meglemosian (11 Kya – 8 Kya) cultures of Europe, to the Caspian culture of North Africa (12 Kya – 8 Kya). In the Americas human cultural development is differently partitioned. A roughly analogous timeframe to the European Mesolithic matches the American Paleoindian or Lithic Period, which runs from 11.5 Kya to 8.5 Kya. Several distinct traditions are identified within period, though only two, the Clovis culture (13.5 Kya–11 Kya) and the Dalton Tradition (8.5 Kya–7.9 Kya) existed within the Atlantic region. Three others, the Western Fluted Point Tradition, the California Post Pattern, and the Folsom tradition all date to the same general period, but were located in the American West. At least some of these Paleoindian peoples were part of what was at least the second wave of human migrants into the Americas. They arrived around 11.2 Kya and moved into a continent with a pre-existing human population. We don't know much about the very first Americans, who probably arrived over 15.5 Kya. There are a few scattered habitations sites, in widely separated locations in both continents, including Texas (Waters et al. 2018), Pennsylvania (Carr 2018), South Carolina (Goodyear 2005), and Southern Chile (Meltzer 1997).

An intriguing characteristic of the Mesolithic is our relatively limited understanding of much of human life in this time. This is in part because of the historical distance from which we view it. More importantly the human predilection for living beside bodies of water, particularly in coastal areas, means that much of the territory inhabited by Greenlandian peoples is now inundated. The rapid rise in sea level throughout the North Atlantic between 11.7 Kya and 8.5 Kya drowned many of the landscapes they occupied. There are exceptions, of course, and some important sites have been identified. However, it is only with the past two decades that detailed geographies of some of the flooded land have been reconstructed. This is primarily because in recent decades detailed bathymetric data from multibeam sonar surveys has become widely available, particularly in those places such as the North Sea, where commercial mineral exploration surveys have been made available to researchers by industrial partners. These data can be used to reconstruct accurate and detailed three-dimensional reconstructions of submerged landscapes built on underwater archaeology, geological sampling, and sediment analysis (Fitch et al. 2007).

This reconstructive work is revealing lost environments on both sides of the Atlantic, in what is now the North Sea, and in the Gulf of Maine. It offers hints of other less well explored territory in the Gulf of Mexico, the Caribbean, along the East Coast of North America, and off the coast of the Canadian Maritimes. Not all of these areas were necessarily home to humans, but they were associated with a tapestry of terrestrial, estuarine, littoral, intertidal, and nearshore shallow marine ecosystems. These are typically highly productive, diverse, and dynamic environments. We know that when humans did find these places, they frequently developed cultural adaptations that exploited a wide variety or the diverse available resources, and created complex, and relatively populous, societies. Then, as now, a large proportion of the human population lived in coastal and estuarine regions, and at lower altitudes. Then, as now these environments are vulnerable to both short and long-term changes. The price paid for enjoying the ecologically productive junction between land and sea is to contend with environments that are geomorphologically very dynamic.

As previously noted, the European landscapes showing the greatest differences from their current geography in the lowlands of the North Sea (Doggerland), the Southern Irish Sea and Solent, the coast of Brittany, the Northern Adriatic, and the Northern Caspian, also changed the most rapidly. As the Greenlandian ended, and they were inundated by rapidly rising sea level. The most thoroughly researched of these, Doggerland, was not a marginal environment, but rather served as a cultural core. Large areas of the landscape comprised coastal, estuarine, and riverine wetlands; the type of landscape often associated with more settled lifeways for gatherer-hunters. This presents an intriguing possibility, outlined by archaeologist Vince Gaffney, that Doggerland may have been the cradle of Northern European agriculture. Samples from Doggerland dated to around 8 Kya contain sedimentary ancient DNA (sedaDNA) from domesticated, cattle, dogs, and wheat. This predates the earliest recorded domestication in Britain by almost 2000 years, suggesting a couple of things. First, that the inhabitants of Doggerland had a cultural network that extended across the rest of Europe to the original sites of wheat domestication in the Levant. Secondly, that the transition from Mesolithic gathering, hunting and fishing, to

Neolithic agriculture was a complicated process, that probably involved "ephemeral pioneer events" (Gaffney et al. 2017). That is, some people periodically experimented with agriculture, and then abandoned it, only for others to try again in another time and place.

In the Americas the cultural shift at the beginning of the Greenlandian is not as clearly demarcated as it is in Europe. Paleoindians were in Atlantic landscapes both during and after the LGM (Lothrop et al. 2011). Like their European contemporaries they experienced an analogous series of changes as ice melted, new land emerged from beneath the glaciers, and also from beneath the ocean. At the lowstand, 11 Kya sea levels were 50 m below their current levels, rising fairly rapidly until around 10 Kya, when they were around 20 m below their current level (Kelley et al. 2010). Thereafter there was a much slower rise ("slowstand") until around 7.5 Kya, when inundation accelerated again. Although there were similarly huge ecological and geographical changes, the beginning of the Greenlandian does not seem to have a clearly identifiable cultural marker in the Americas in the same way that it does in Europe. It may well be that researchers' lack of access to widespread industry bathymetric and coring survey data for the American East Coast means that we simply know a lot less about how Paleoindians adapted to the end of glaciation, and the opening of new landscapes. We can reasonably hypothesize that they must have developed new survival strategies, which must in turn have required changes in culture and technology. But the biggest changes identified in the record to date have predated the end of the LGM. The Clovis people were producing finely worked tools –a diagnostic feature of the Neolithic Europe – before the end of the Greenlandian got underway.

There are tantalizing indications that the now submerged early Holocene landscapes of America was home to cultures that are as yet little known. As in Europe there have been a steady stream of artifacts retrieved by commercial fishermen, particularly stone tools and occasionally human or animal remains. However, absent the systematic and broad survey available to European researchers, there are gaps in our knowledge. Where surveys have been undertaken, there are indications that Greenlandian Paleoindians also occupied coastal and estuarine landscapes. Presumably like their European brethren the rich and varied biological resources of coastal wetlands and estuaries supported a good lifestyle. Near Bass Harbor, on Mount Desert Island, Maine there is a submerged site that has been well surveyed. Kelley, Belknap, Kelley and Claesson have been able to model the prehistoric landscape of this area. They suggest that artifacts retrieved by scallop draggers in the area came from a wetland, estuary and beach complex. They have dated the site to some time between 10.3 Kya and 8 Kya BP (Kelley et al. 2013).

Both our more detailed understanding of the European case, and the intriguing possibilities presented by our more fragmented understanding of the American situation, leave us with an incomplete picture. Broadly stated, the steadily increasing variety of Mesolithic cultures indicates that there were complex and entirely novel human social and technological processes at work throughout this period. The net result was a significant increase in the number of people in the world, an infilling of habitable niches by humans, and a consequent explosion of cultural diversity.

One cultural adaptation of particular importance that started to emerge in the Fertile Crescent of Iran, Syria, Iraq, and the Levant was the beginning of the long process of domestication. The pre-pottery Neolithic sites in the Fertile Crescent show a transition from early wild plant management to domesticated species in the period between 10 Kya and 8.5 Kya (Riehl et al. 2013). While some people were busy exploring the potential of domesticating wild plants and animals, others were exploring the possibilities of ceramics: pottery seems to have initially been a hunter-gatherer innovation (Craig et al. 2013). Within the same timeframe the first indications of long-term settlements emerged in a number of places (Anderson et al. 2007). That is, while all occurring within the timeframe of the Greenlandian, it seems that agriculture, ceramics, and permanent settlement were all decoupled processes. The impression we have is of a dynamic physical environment and a matching dynamic cultural environment, with unprecedented movement and change happening in both spheres.

3.6 The End of the Greenlandian

The end of the Greenlandian is marked by the 8.2 k event, over a century and a half of cooling that occurred around the eponymous date. Indicators of this cooling have been retrieved across the globe, from the Arctic Greenland to Southern Africa. Cores from ice caps, ocean sediments, peat bogs, and lakes have yielded oxygen isotopes, plankton, and pollen. Stalactites and stalagmites have delivered speleotherm records. The fossilized middens of cave-dwelling rock hyrax have yielded pollen and isotopes. The broad pattern is a world-wide temperature decline of between 1 °C and 5 °C. This started around 8175 BP, with a drop of over 3° in the first two decades. Following this there was a 60-year cold period, and then after about 150 years, a recovery (Walker et al. 2019) (Fig. 3.7).

The prevailing theory is that the 8.2 k event was triggered by the final collapse of the Laurentide Ice Sheet, and the release of cold, fresh water from two of its associated lakes, Agassiz and Ojibway. This slug of cold water may have been released in a rather dramatic fashion, in either one or two major catastrophic failures (Godbout et al. 2020). Calculations suggest that during the flood may have comprised "5 million cubic metres per second, flowing for a period of about six months" – which is a lot of water, for a long time (Jakobsson 2008). It was sufficient to displace the flow of the Gulf Stream, and so redirect warm tropical water southwards, triggering over a century and a half of cooling in the high latitude North Atlantic (Fig. 3.8).

In what can at the very least be described as an extremely unfortunate coincidence for everything living in the North Atlantic during the Greenlandian, the catastrophic discharge of lakes Ojibway and Agassiz and the consequent changes in climate and sea level happened in the same time frame an earthquake off the coast of Norway which triggered the Storegga Slide, a submarine landslide that set off the Storegga Tsunami. The Storegga Slide probably shared some common causation with the collapse of North American glacial lakes as it may have resulted from the

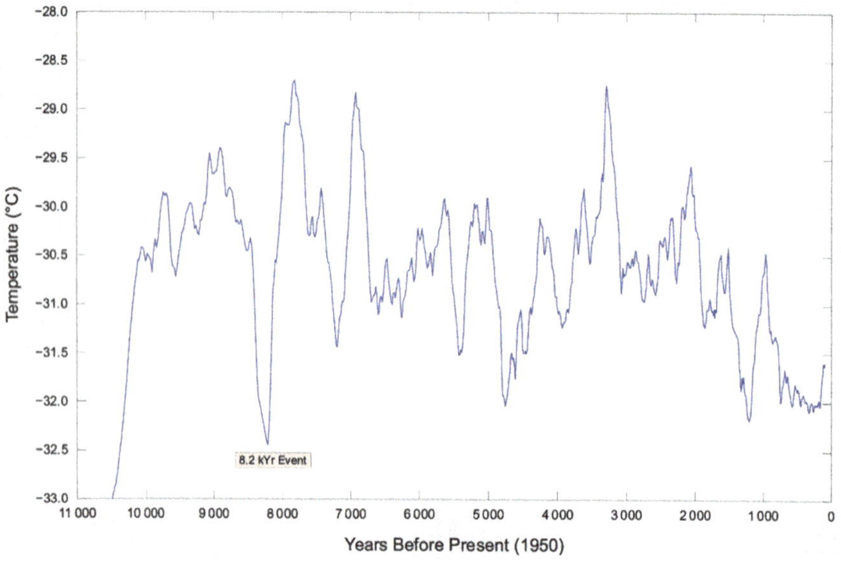

Fig. 3.7 The global cooling of the 8.2 kilo event seen in the Greenland Ice Core for which it is named

final effects of the collapsing ice sheets. In the cases of Storegga, the mid-Norwegian continental shelf was prone to this type of megaslide, having experienced several similar events over the past 2.5 million years. These past events all appear to be associated with cycles of glaciation. In the case of Storegga, and probably its predecessors, the sequence of events started with an accumulation of proglacial sediments during an ice-age. As the glaciers receded there was an accompanying accumulation of methane gas hydrate-bearing sediments (clathrates), causing upwards pressure on the accumulated materials. Earthquake activity resulting from the continental rebound following the removal of the LGM ice burden provided the final impetus for the slide. The Storegga Slide is the largest known landslide on Earth, and it generated a tsunami that that ran 10–12 m up the west coast of Norway, 4–6 m in Scotland, 20–30 m in Shetland, and 10 m in the Faroes (Bryn et al. 2005).

These two events contemporaneous events in the mid-latitude North Atlantic appear to have shared a common trigger in the final collapse of the relict ice-sheets of the LGM. Both were catastrophic, and both culminated in Biblical-scale inundations of surrounding lands. In the case of the British Isles, there was a significant population decline as fertile coastal lowlands were abandoned, and climate and ecosystems were transformed (Waddington and Wicks 2017). We can reasonably assume that comparable effects occurred on the western side of the Atlantic.

To summarize, the Greenlandian ended with a series of catastrophic floods, a permanent loss of coastal lowlands, and over a century of destructive cold as ocean currents were disrupted.

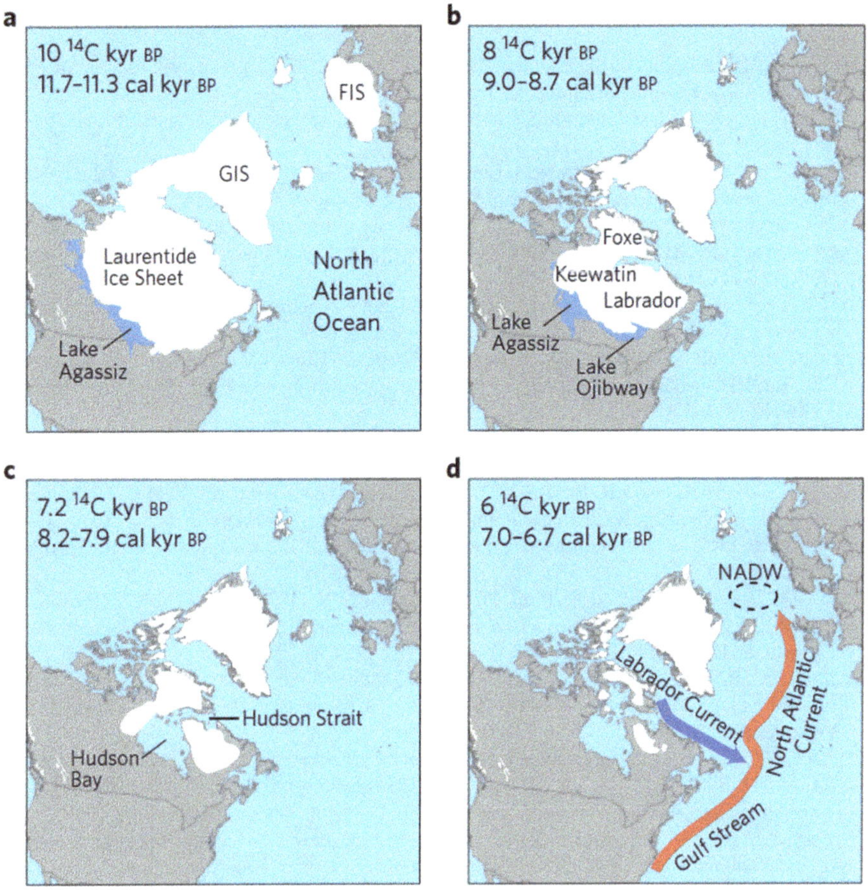

Fig. 3.8 Glacial Lakes Ojibway and Agassiz, and their influence on the Gulf Stream (Törnqvist and Hijma 2012)

References

Anderson DG, Russo M, Sassaman KE (2007) Mid-Holocene cultural dynamics in southeastern North America. In: Climate change and cultural dynamics. Academic, Amsterdam, pp 457–489

Araujo BB, Oliveira-Santos LG, Lima-Ribeiro MS, Diniz-Filho JA, Fernandez FA (2017) Bigger kill than chill: the uneven roles of humans and climate on late quaternary megafaunal extinctions. Quat Int 28(431):216–222

Ayache M, Swingedouw D, Mary Y, Eynaud F, Colin C (2018) Multi-centennial variability of the AMOC over the Holocene: a new reconstruction based on multiple proxy-derived SST records. Glob Planet Chang 170:172–189

Barnosky AD, Lindsey EL, Villavicencio NA, Bostelmann E, Hadly EA, Wanket J, Marshall CR (2016) Variable impact of late-quaternary megafaunal extinction in causing ecological state shifts in North and South America. Proc Natl Acad Sci 113(4):856–861

Bond DP, Grasby SE (2017) On the causes of mass extinctions. Palaeogeogr Palaeoclimatol Palaeoecol 478:3–29

Bryn PK, Berg K, Forsberg F, Solheim A, Kvalstad TJ (2005) Explaining the Storegga slide. Mar Pet Geol 22:11–19
Carr KW (2018) Peopling of the middle Atlantic. In: Middle Atlantic prehistory: foundations and practice, vol 5. Rowman & Littlefield, Lanham, p 219
Chen X, Tung KK (2018) Global surface warming enhanced by weak Atlantic overturning circulation. Nature 559(7714):387–391
Cléroux C, de Menocal P, Arbuszewski J, Linsley B (2013) Reconstructing the upper water column thermal structure in the Atlantic Ocean. Paleoceanography 28(3):503–516
Colin C, Frank N, Copard K, Douville E (2010) Neodymium isotopic composition of deep-sea corals from the NE Atlantic: implications for past hydrological changes during the Holocene. Quat Sci Rev 29(19–20):2509–2517
Craig OE, Saul H, Lucquin A, Nishida Y, Taché K, Clarke L, Thompson A, Altoft DT, Uchiyama J, Ajimoto M, Gibbs K (2013) Earliest evidence for the use of pottery. Nature 496(7445):351–354
Estes JA, Terborgh J, Brashares JS, Power ME, Berger J, Bond WJ, Carpenter SR, Essington TE, Holt RD, Jackson JB, Marquis RJ (2011) Trophic downgrading of planet Earth. Science 333(6040):301–306
Fitch S, Gaffney V, Thomson K (2007) In sight of doggerland: from speculative survey to landscape exploration. Internet Archaeol 22(2007):1–10
Gaffney V, Allaby R, Bates R, Bates M, Ch'ng E, Fitch S, Garwood P, Momber G, Murgatroyd P, Pallen M, Ramsey E (2017) Doggerland and the Lost Frontiers Project (2015–2020). In: Under the sea: archaeology and Palaeolandscapes of the continental shelf. Springer, Cham, pp 305–319
Galetti M, Moleón M, Jordano P, Pires MM, Guimaraes PR Jr, Pape T, Nichols E, Hansen D, Olesen JM, Munk M, de Mattos JS (2018) Ecological and evolutionary legacy of megafauna extinctions. Biol Rev 93(2):845–862
Godbout PM, Roy M, Veillette JJ (2020) A detailed Lake-level reconstruction shows evidence for two abrupt lake drawdowns in the late-stage history of the eastern lake Agassiz-Ojibway basin. Quat Sci Rev 238:1–13
Goodyear AC (2005) Evidence of pre-Clovis sites in the eastern United States. In: Bonnichsen R, Lepper BT, Stanford D, Waters MR (eds) Paleoamerican origins: beyond Clovis. University of Texas Press, pp 103–112
Gregoire LJ, Ivanovic RF, Maycock AC, Valdes PJ, Stevenson S (2018) Holocene lowering of the Laurentide ice sheet affects North Atlantic gyre circulation and climate. Clim Dyn 51(9–10):3797–3813
Hays JD, Imbrie J, Shackleton NJ (1976) Variations in the Earth's orbit: pacemaker of the ice ages. Science 194:1121–1132
Jakobsson M (2008) The last stampede of a glacial lake. Nat Geosci 1(3):152–153
Jaunsproge M (2013) Cross-media 3D cartography of 'Europe at the last ice age' based on initial data compilations. MA Thesis. Fakultät Umweltwissenschaften. Technische Universität Dresden
Kaufman DS, Ager TA, Anderson NJ, Anderson PM, Andrews JT, Bartlein PJ, Brubaker LB, Coats LL, Cwynar LC, Duvall ML, Dyke AS (2004) Holocene thermal maximum in the western Arctic (0–180 W). Quat Sci Rev 23(5–6):529–560
Kelley JT, Belknap DF, Claesson S (2010) Drowned coastal deposits with associated archaeological remains from a sea-level "slowstand": Northwestern Gulf of Maine, USA. Geology 38(8):695–698
Kelley JT, Belknap DF, Kelley AR, Claesson SH (2013) A model for drowned terrestrial habitats with associated archeological remains in the Northwestern Gulf of Maine, USA. Mar Geol 338:1–6
Khan NS, Ashe E, Shaw TA, Vacchi M, Walker J, Peltier WR, Kopp RE, Horton BP (2015) Holocene relative sea-level changes from near-, intermediate-, and far-field locations. Curr Clim Chang Rep 1(4):247–262
Koul V, Tesdal JE, Bersch M, Hátún H, Brune S, Borchert L, Haak H, Schrum C, Baehr J (2020) Unraveling the choice of the north Atlantic subpolar gyre index. Sci Rep 10(1):1–2

Lochte AA, Repschläger J, Kienast M, Garbe-Schönberg D, Andersen N, Hamann C, Schneider R (2019) Labrador Sea freshening at 8.5 ka BP caused by Hudson Bay Ice Saddle collapse. Nat Commun 10(1):1–9

Lothrop JC, Newby PE, Spiess AE, Bradley JW (2011) Paleoindians and the younger Dryas in the New England-Maritimes region. Quat Int 242(2):546–569

Lövei GL (2007) Extinctions, modern examples of. In: Levin SA (ed) Encyclopedia of biodiversity. Elsevier, Amsterdam, pp 1–13

Lyons SK, Smith FA, Wagner PJ, White EP, Brown JH (2004) Was a 'hyperdisease' responsible for the late Pleistocene megafaunal extinction? Ecol Lett 7(9):859–868

Mann DH, Groves P, Gaglioti BV, Shapiro BA (2019) Climate-driven ecological stability as a globally shared cause of late quaternary megafaunal extinctions: the plaids and stripes hypothesis. Biol Rev 94(1):328–352

Marcott SA, Shakun JD, Clark PU, Mix AC (2013) A reconstruction of regional and global temperature for the past 11,300 years. Science 339(6124):1198–1201

McCormick SD, Regish AM, Ardren WR, Björnsson BT, Bernier NJ (2019) The evolutionary consequences for seawater performance and its hormonal control when anadromous Atlantic salmon become landlocked. Sci Rep 9(1):1–0

Meltzer DJ (1997) Monte Verde and the Pleistocene peopling of the Americas. Science 276(5313):754–755

O'Regan M, Williams CJ, Frey KE, Jakobsson M (2011) A synthesis of the long-term paleoclimatic evolution of the Arctic. Oceanography 24(3):66–80

Pershing AJ, Christensen LB, Record NR, Sherwood GD, Stetson PB (2010) The impact of whaling on the ocean carbon cycle: why bigger was better. PLoS One 5(8):e12444

Pielou EC (1991) After the ice age. University of Chicago Press, Chicago

Riehl S, Zeidi M, Conard NJ (2013) Emergence of agriculture in the foothills of the Zagros Mountains of Iran. Science 341(6141):65–67

Saini J, Stein R, Fahl K, Weiser J, Hebbeln D, Hillaire-Marcel C, de Vernal A (2020) Holocene variability in sea ice and primary productivity in the northeastern Baffin Bay. Arktos 25:1–9

Shakun JD, Clark PU, He F, Marcott SA, Mix AC, Liu Z, Otto-Bliesner B, Schmittner A, Bard E (2012) Global warming preceded by increasing carbon dioxide concentrations during the last deglaciation. Nature 484(7392):49–54

Shapiro RS, Spangler E (2009) Bacterial fossil record in whale-falls: petrographic evidence of microbial sulfate reduction. Palaeogeogr Palaeoclimatol Palaeoecol 274(3–4):196–203

Smith CR, Glover AG, Treude T, Higgs ND, Amon DJ (2015) Whale-fall ecosystems: recent insights into ecology, paleoecology, and evolution Marine. Science 7:572–596

Su Z, Ingersoll AP, He F (2016) On the abruptness of Bølling–Allerød warming. J Clim 13:4965–4975

Sumida PY, Alfaro-Lucas JM, Shimabukuro M, Kitazato H, Perez JA, Soares-Gomes A, Toyofuku T, Lima AO, Ara K, Fujiwara Y (2016) Deep-sea whale fall fauna from the Atlantic resembles that of the Pacific Ocean. Sci Rep 6(1):1–9

Thiagarajan N, Gerlach D, Roberts ML, Burke A, McNichol A, Jenkins WJ, Subhas AV, Thresher RE, Adkins JF (2013) Movement of deep-sea coral populations on climatic timescales. Paleoceanography 28(2):227–236

Thiagarajan N, Subhas AV, Southon JR, Eiler JM, Adkins JF (2014) Abrupt pre-Bølling–Allerød warming and circulation changes in the deep ocean. Nature 511(7507):75–78

Törnqvist TE, Hijma MP (2012) Links between early Holocene ice-sheet decay, sea-level rise and abrupt climate change. Nat Geosci 5(9):601–606

Van der Bilt WG, DAndrea WJ, Werner JP, Bakke J (2019) Early Holocene temperature oscillations exceed amplitude of observed and projected warming in Svalbard lakes. Geophys Res Lett 46(24):14732–14741

Waddington C, Wicks K (2017) Resilience or wipe out? Evaluating the convergent impacts of the 8.2 ka event and Storegga tsunami on the Mesolithic of Northeast Britain. J Archaeol Sci Rep 14:692–714

Walker M, Head MJ, Lowe J, Berkelhammer M, Björck S, Cheng H, Cwynar LC, Fisher D, Gkinis V, Long A, Newnham R (2019) Subdividing the Holocene series/epoch: formalization of stages/ages and subseries/subepochs, and designation of GSSPs and auxiliary stratotypes. J Quat Sci 34(3):173–186

Waters MR, Keene JL, Forman SL, Prewitt ER, Carlson DL, Wiederhold JE (2018) Pre-Clovis projectile points at the Debra L. Friedkin site, Texas – Implications for the Late Pleistocene peopling of the Americas. Sci Adv 4(10):1–13

Williams MA, Dunkerley DL, De Deckker P, Kershaw AP, Stokes TJ (1993) Quaternary environments, 2nd edn. Edward Arnold Publishers Ltd, London

Williams JW, Shuman BN, Webb T III, Bartlein PJ, Leduc PL (2004) Late-quaternary vegetation dynamics in North America: scaling from taxa to biomes. Ecol Monogr 74(2):309–334

Zobel M, Davison J, Edwards ME, Brochmann C, Coissac E, Taberlet P, Willerslev E, Moora M (2018) Ancient environmental DNA reveals shifts in dominant mutualisms during the late quaternary. Nat Commun 9(1):1–9

Chapter 4
The Northgrippian

The Time of Transmutation (8200 – 4,200 BP)

Abstract The second of the Holocene's three ages. The Northgrippian temperature plateau bracketed by two major climate disruptions, the 8.2 k event, and the 4.2 k event. The four transformative technological and cultural developments that became widespread, often in combination, during the comparative warmth and stability of the Northgrippian: ceramics, domestication, villages and agroecosystems, and smelting.

Keywords Ceramics · Domestication · Villages · Agroecosystems · Smelting · 4.2 k event

4.1 Introduction

The second stage of the Holocene, the Northgrippian is defined by a marker in the NGRIP1 Greenland ice core, for which it is named (Walker et al. 2019). It is bracketed by two cooling events, the brief but dramatic 8.2 k event, the hiccup in the overall warming trend of the first half of the Holocene, and the 4.2 k event, which was a longer directional change, when climate patterns shifted and stayed shifted. The Northgrippian represents the middle of the ζ curve and was a time of generally warmer temperatures globally – it has been called a "temperature plateau" (Marcott et al. 2013), though it was, if fact, more of a moderate arc.

The Northgrippian is the earliest time from which we have clear human voices. In the NAPT there are languages, stories, legends, religious texts, and art that form a continuous tradition reaching over 8000 years to the present: the end of the Neolithic is within our collective memory. For example, the early books of the Old Testament describe a pre-metallurgical nomadic herding society. Native American stories describe the replacement of Archaic peoples by Early Woodland people. Norse tales reference pre-metal cultures. And so on.

During the Northgrippian as physical geographies stabilized into their current configurations ecosystems and human cultures underwent dramatic and consequential changes. Central to both of these changes was the establishment of enduring

M. Bampton, *The North Atlantic Polar Triangle*, Springer Polar Sciences,
https://doi.org/10.1007/978-3-031-27264-6_4

agroecosystems, first in the Fertile Crescent region of the Eastern Mediterranean, and subsequently in the rest of Europe, Africa, and the Americas. The central features of agroecosystems are long-term manipulations of landscapes by multigeneration genetic engineering of plants and animals. This accompanied by a matching engineering of soil structures and hydrology. The widespread adoption of either one or both of two other major technologies accompanies the development of agroecosystems: ceramics and metallurgy.

All of these changes have origins that can be traced back to the Greenlandian, and most were developed independently in several locations. However, they became established and spread outwards from isolated centers of innovation, combining and cross-pollinating with one another, during the Northgrippian. Why this should have happened at this particular point in history deserves some consideration. There are several candidate causal circumstances. The transformed post-glacial climate provided a situation of comparatively benign weather. This was coupled with the continuing loss of many large mammal prey species and (also) a few predators. And it was the first time that anatomically modern humans had faced both the opportunities and the challenges posed by these new conditions in places where existing landscapes contained resources suited for these developments.

The Northgrippian closely matches the Eastern Hemisphere Neolithic. In the Americas the division is less clearly demarcated, but it roughly corresponds to the beginning of the Archaic Period, which is typically identified as running from 8000 BP – 1000 BP. The periodization of cultural developments in Africa is less clearly defined. However, during this same period agriculture, ceramics and metals were developed in West Africa (beyond the scope of the present discussion, similar developments occurred in Asia).

In each of these cases the technological, cultural and ecological innovations happened independently from one another. When they emerged, the new developments did not always prosper and did not play out in the same ways in each place. But it is unavoidable to see the consistent patterns in space and time represented by these developments, particularly when geographical diffusion resulted in their combining with one another. In this chapter, I shall organize this discussion into three major components. First, an overview of the climate dynamics. Secondly a discussion of the impacts of climate on geography and oceanography. Thirdly, a discussion of these changes on ecology and culture. These last two are now combined, in contrast to the previous chapter, as they are inextricable entwined at this stage.

This last component of the discussion is particularly important, because it is in the Northgrippian that the distinctions between human and non-human forcing agents starts to invert. We have already seen that some processes, such as regional megafaunal extinctions, are inextricably linked to the arrival of humans in landscapes during the Greenlandian. This was particularly notable in the Americas and in Australia, but it also happened elsewhere at other times in history. Following the arrival of humans Australia, Madagascar, and New Zealand all saw catastrophic declines and eventual extinctions of the majority of their native megafauna.

However, in these cases there were other dynamics beyond extinctions that were significant players in the development of local ecosystems. And there were ecosystems that remained untouched by humans. In the Northgrippian the balance slowly started to tip from non-human forcing towards human forcing. This is most immediately evident when one realizes that there are fewer processes and fewer systems worldwide that were untouched by people, either directly or indirectly. As human populations grew, as people moved into new environments, and as agriculture became the means of production for a higher proportion of the human race, there were more cases in which there was competition between humans and other species for resources. And there were increasingly fierce competitions between people and other people for real estate. Landscapes became contested in ways previously unknown.

This does not negate the operation of the non-human forcing agents, of course. The historical dynamics of non-human natural processes continued. Milankovitch cycles still operated, volcanoes still erupted, and oceans and ice sheets still functioned. Within ecosystems there were still plant and animal evolutionary dynamics operating. Sometimes these worked in response to humans, sometimes in parallel with humans, and sometimes completely independently of humans. There were also entire ecosystems almost completely untouched by human action in any form. Thus, ocean ecosystems, particularly those in deep oceans, and remote islands, such as those of Oceana, all remained isolated and untouched.

This noted, the Northgrippian marks a qualitative change in the role humans played in the operation of Earth systems. This change was driven by a set of objective and exogenous forces, in the first instance. The climate conditions that compelled humans to change their survival strategies, and ultimately provided the opportunity for the spread of agriculture was one such exogenous factor. Similarly, the ecological and geographical conditions that favored and supported the development of agriculture were also factors. Understanding the Northgrippian requires and understanding of this tension between the human and the non-human, and how it played out prior to the 4,.2 k event, which shifted the dynamic once again and ended the Northgrippian.

The shift from predominantly village-based agriculture towards more populous, technologically sophisticated, and hierarchical human cultures was well established by the end of the Northgrippian. Why this occurred during this time, and what drove the processes is a question I will sidestep for the present. As the middle third of the Holocene, the "belly" of the ζ-curve, passed a new phase of climatic, physical, and human processes emerged. Although not as clearly demarcated as some other chronostratigraphic shifts, within the NAPT there is clearly a "before" and an "after" to this change. I will deal with the "after" in the next chapter, for now I shall focus on the most important characteristics of climate, geography, ecology and culture.

4.2 Climate

The variation in latitudinal climate patterns that characterized the Greenlandian was less apparent through the Northgrippian, but there was still a clearly defined spatial organization to changes. Despite geographical, centennial and millennial fluctuations around a set of mean values and trends, this was a period in which for roughly 4000 years climate condition remained fairly stable worldwide (Fig. 4.1 Northgrippian climate). The temperature trend held from the poles to the equator and was reflected in similarly stable patterns of precipitation. While there were, of course, substantial variations, a global mean of about 0.4° above the series average held throughout. The primary driver of this thermal maximum was astronomical, the result of the combination of earth-sun geometric changes that we collectively lump into the term Milankovitch forcing. The two most important Milankovitch forcing agents during this time were an increased axial tilt of 0.5° and the coincidence of the perihelion, the time at which Earth's orbit brings it closest to the Sun, with the Northern Hemisphere's summer.

The increased axial tilt, from 23.5° to 24°, enlarged both Arctic circles, pushing them half a degree Equatorwards. A rough calculation indicates that given the Earth's radius at the pole is 6357 the circumference of the current Arctic Circle at 66.5° is around 2535 km. Its area, the Earth's Arctic spherical cap, is 21,059,494 km^2. Increasing the tilt of the Earth's axis by 0.5° extends the Arctic Circle by 0.5° to 66° and increases its radius to 2586 km. This increases the area of the Arctic spherical cap to 21,951,900 km^2, a gain of 892,406 km^2. In total this yields an increased Arctic land area of 1,784,812 km^2 divided between the two poles. The climatic and ecological consequences of this are, to put it mildly, significant. The decrease in Northern Hemisphere insolation during the winter months, and the concomitant increase during the summer months served to increase the latitudinal energy

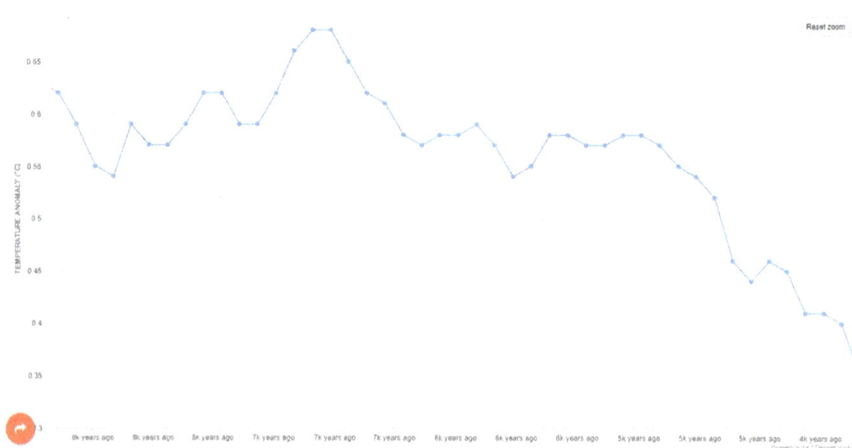

Fig. 4.1 Global average temperature in the Northgrippian (After https://www.temperaturerecord.org/)

gradient within each hemisphere, and the seasonal energy differential between hemispheres. In other words, there were darker colder high latitude winters, over a greater area and for a longer time. However, the effects of this were offset by increased day length and more solar radiation input during the summers.

The Earth's orbit around the Sun currently has an eccentricity of around 0.0167. This translates to a variation of around 6% in incoming solar radiation between the perihelion and the aphelion, the closest and farthest points from the Sun respectively. Whenever the perihelion coincides with one of the hemispheric summers the increased insolation causes an increase in yearly average temperature. In other words, the orbital eccentricity, in combination with the increased summer radiation input from the expanded Arctic Circle, offset the effects of the increased albedo and extended winter night resulting from extended winter ice cover. The net result of the increased axial tilt and the perihelion effect resulted in 0.2% more solar radiation, an increase of between 28 and 36 W/m^2, in the Northern Hemisphere in summer. This was sufficient to trigger a series of feedback loops and coupled effects to produce the Holocene Climatic Optimum (HCO), with a warming trend that peaked around 8 Kya, shortly before the 8.2 k event triggered the beginning of the Northgrippian. After the 8.2 k event, astronomically driven high temperatures continued for the duration of the Northgrippian.

The Northern Hemisphere temperature increase resulted in a significant reduction in sea ice. Coverage was comparable to the historical low now we are now experiencing. Then, as now, by lowering albedo this has a significantly amplified the effects of increased incoming solar radiation. The resulting southward shift of the intertropical convergence zone, the belt of tropical winds and rains defining the humid tropics. This was accompanied by weaker monsoons over Africa and Asia. Dryness and desertification extended into lower latitudes on both sides of the Atlantic, and at least in parts of the Americas with some ancillary effects, such as an increase in wildfires (Mendieta et al. 2018).

Continued Milankovitch forcing over the course of the Northgrippian gradually caused a redistribution of solar energy, on a millennial timescale, progressively shifting the Northern Hemisphere summer extent of the Intertropical Convergence Zone (ITCZ). By 4.2 Kya this resulted in sufficient cooling to usher in the third part of the Holocene, the Meghalayan. The signal for the 4.2 k event is much less clearly defined, and significant debates about its legitimacy remain active. There are mixed physical signals from place to place, and there is no clear single indication of a global catastrophic event. However, there were a series of aridity events dispersed around the Atlantic.

Ocean circulation through this period was influenced by the enhanced energy difference between low and high latitudes. The North Atlantic Oscillation (NAO) progressively shifted into a more negative pattern, and the AMOC was disrupted. Changes in vegetation driven first by regional warming and later by human action, coupled with changes in oceanic temperature, sea-ice cover, and oceanic circulation, resulted in a positive feedback loop, that by around 6 Kya amplified the overall system response.

4.3 Geography

Northgrippian geographical dynamics were substantially different from those of the Greenlandian. There were no dramatic changes of coastline, with entire regions being reshaped, and no landscapes were lost. No great ice-sheet collapses occurred, and there were no major transformative flood events. Yet over four thousand years incremental process did change things. A significant part of this is attributable to human action, more of which in the next section. The transformation of landscapes during this period marks the change from low-density gatherer-hunter lifeways, interspersed with early village agriculture to widespread agroecosystems (Kalis et al. 2003). There were also a series of changes that were driven primarily by non-human processes. The long-term incremental processes of geomorphology such as weathering, mass movement, erosion, and deposition changed the form and distribution of inorganic materials. This in turn provided nutrients the raw materials for soil formation. Migrating meanders shifted the courses of rivers. Sediment entrainment and deposition budgets shifted in response to changing climate conditions, and the changes in available weathered materials. Consequently, stream profiles changed, estuaries filled, and deltas grew. The changed geographies necessarily influenced the distribution and arrangement of biogeographies. On a map of the world such developments are not immediately apparent. However, on the scale of the lived experience of humans, and the behaviors of individual plants and animals, these changes were profound.

One way to explore the importance of these changes is to use the perspective of climatic geomorphology, as propounded by, among others, Julius Büdel (1973), and Tricart and Cailleux (1972). There is a close link between physical landscapes and prevailing climate conditions. Landscapes contain a signature of the process regime that has shaped them rooted in a much deeper time frame than the daily and yearly cycles of weather. Thus, places above latitude 40° in both hemispheres have comparatively thin soil horizons, since soils, and the weathered parent material from which they are derived, have only be in situ since the beginning of the Holocene. By comparison, the weathering horizons and soils of the humid tropics are extraordinarily deep, as they have been in place for most of the Pleistocene, and a significant proportion of the Pliocene. The climatic geomorphology perspective views any current landscape as a snapshot of its entire history. This is useful in the context of the slowly changing landscapes of the middle Holocene. The comparatively gradual pace of change during this time, and the absence of a clear "line in the sand" makes the incremental transformation of landscapes a useful organizing structure.

4.4 Ecology and Culture

Broadly stated the ecology of the Northgrippian was characterized by the establishment of climax ecosystems in the deglaciated landscapes on both sides of the Atlantic. In the period that Pielou calls "The Great Warmth" she describes the situation in North America as follows:

> A snapshot of conditions late in the hypsithermal in the Great Lakes region is given by fossil finds along the north shore of Lake Ontario that date from between 5 k and 6 k B.P. They suggest a moist, rich forest of beech, sugar maple, and eastern white pine with grassy openings. The diversified habitat was home to a variety of small mammals—chipmunks, squirrels, flying squirrels, deer mice, meadow voles, and shrews and, a step up the food chain, red and grey foxes. There were also ponds and lakes inhabited by muskrats, wood ducks, and red winged blackbirds. The contrast with conditions a few millennia earlier could hardly be greater … Fossil finds show that the geographical ranges of numerous animal species extended farther north in the hypsithermal than they do today. These fossils provide the evidence that there was a, indeed, a warm period in the past and show also how the warmth slowly drifted from west to east across the continent. (Pielou 1991: 271–272)

In lower latitudes climate changes reduced humidity. The African Humid Period, a climate episode in which the Sahara was not a desert, from around 14.6 Kya forwards, drew to a close 5.5 Kya. Prevailing lower latitude aridity is also recorded for the Americas (Schaney et al. 2021). Such differences are important, but of greater significance is the increasing impact of purposeful human interaction with the ecosystem.

Humans of all species have changed and managed landscapes throughout their two-million-year history. The conscious manipulation of the surrounding world is one of the defining characteristics of humanity. However, the Northgrippian saw both quantitative and qualitative changes in the ways humans transformed nature. These transformational behaviors that had previously existed in isolated pockets of space and time became widespread, and commonplace, and some entirely new behaviors evolved. There are four defining examples of this, which I list here in roughly chronological order. First, ceramics, and secondly agriculture. These had both been developed in the Greenlandian, but only became widespread in the Northgrippian. Thirdly, closely coupled with these technological changes, and also previously present in isolated places, were a collection of ecosystem and human social changes that can be lumped (inelegantly) into the broad terms "agroecosystems" and "villages". Fourthly, and entirely without precedent, was the development of smelting.

Unravelling the threads of these four intertwined changes forces us to look outside the NAPT, and before the Northgrippian. The patterns of innovation and diffusion, of the four developments are uneven in both space and time, as mentioned in the previous chapter, reaching back almost to the beginning of the Greenlandian. They occurred in the context of an increasingly complex human demographic picture. Human population increased, from an estimated global total of around two million souls at the beginning of the Greenlandian, to around five million at the beginning of the Northgrippian, rising to around 11 million by its end (Klein et al. 2011). Northgrippian people were spread over a much larger area than at any previous time in history, inhabiting every continent except Antarctica, and a much greater variety of ecosystems. There were numerous culturally and genetically distinct populations that were geographically isolated for long periods. In some places population densities significantly increased. This was particularly so in the most fertile and benign regions such as the eastern end of the Mediterranean and the river valleys of North America (Anderson et al. 2007). The population expansion was matched by increasing economic, linguistic and genetic diversity. On the eastern side of the Atlantic cultures in the first part of the Northgrippian are commonly described using the catch-all phrase "Neolithic", for the New Stone Age. However, there are numerous subdivisions with this broad definition. On the Western side of the Atlantic the same time period roughly corelates with the beginning of the Middle Archaic Period. It is similarly associated with the development of ceramics, agriculture, metallurgy, and villages. There are, of course, significant historical, cultural, and technological differences between various parts of the NAPT. What is striking are the broad similarities that emerged between widely separated locations within the same timeframe. One way to explore this is to look at the three scales at which the changes were made. Through this lens humans' novel manipulation of nature occurred at a molecular scale, at an organism scale, and at an ecosystem scale.

On the molecular level three transmutational technologies formed the core of the trend: ceramics, smelting, and domestication. Each of these three types of manipulation of nature relied on fundamentally changing the molecular character of the things transformed. Ceramics and smelting relied on heat to purify and reform inorganic materials. Domestication relied on the manipulation of DNA. This translated into organism-scale transformations as humans, plants and animals were all changed by domestication. The consequences of molecular and organism-scale changes propagated into systemic change with the development of agroecosystems around the settlements supported by domesticated species. The changes also propagated outwards into other areas of human economy, by allowing for new types of interaction with the natural environment. Ceramic vessels expanded options of food preparation and storage. Metal tools allowed from the development of new types of work, and faster work. Novel agroecosystems allowed for new types of interactions with the environment.

In addition to the planned outcomes, the transmutational changes had a growing number of unintended consequences for humans, plants, and animals. There were necessary changes in human metabolic processes to cope with new foodstuffs. These were accompanied by transformations of human pathologies, as viruses and bacteria moved from animals to humans, and between increasingly densely packed human populations. Human fertility rates changed, as sedentism removed the need for people to carry infants while travelling. Human and animal behaviors were changed, both intentionally by the processes of domestication, and unintentionally by the demands of urban living. Enumerating and tracking these changes, even if we restrict our thinking to the NAPT, is a monumental undertaking still not yet even close to completion. However, we can make a broad generalization about this period of history. It was the time in which the genetic ascendency of sedentary agricultural humans accelerated onto its current trajectory. I mean this in purely biological terms. As many scholars have pointed out, by some well-documented objective measures, the lives of individuals became significantly less successful. Life expectancy decreased. Stature decreased. Dietary diversity decreased. Pathologies increased. Infant mortality increased. Workload increased. Conflict increased. Death-rates increased. Yet despite periodic catastrophic collapses the total number of humans rose (Harari 2018). Once domestication and agriculture were established farming communities grew outwards to occupy every viable ecosystem. They consistently pushed hunter-gatherer-fisher communities into marginal environments. Although some of these communities survived through to the present day, they only do so in the remote and challenging locations, where agriculture is difficult or impossible. On the most ingenious, robust, fortunate, and isolated pre-agricultural peoples preserved the old ways of living. For the rest, once drawn into the agricultural vortex, the die was cast until the next revolution in production.

4.4.1 *Ceramics*

The oldest known ceramics, and therefore the oldest know human-manufactured synthetic objects are the "Venus figurines" of the Gravettian culture, from 30,000 BP. However, ceramic vessels are a more recent development. The oldest known examples are from China, and possibly date from 18,000 BP (Medvedev and Tsetlin 2013). Within the NAPT the first ceramic vessels were made in Mali, around 11,400 BP, and in Amazonia around 7600 BP. Ceramic vessels, rather than decorative or sacred objects were independently developed somewhat later, by a number of cultures within the NAPT. The timing of this development is related to the development if sedentism and agriculture, though in some cases there was a significant gap in time between these events.

Ceramic vessels are a transformational technology for individuals. Pottery allows people to carry and store liquids and other awkward to transport things such as powders, grains and seeds. They opened the way for a whole range of other molecular-level manipulations of nature, from cooking, through pickling and salting, to fermentation. These expanded nutritional opportunities by making food easier to digest and preserve. While all of these things were possible in pre-pottery cultures, the relationship of effort to outcome is quantitatively and qualitatively changed. Increasing technical sophistication in pottery-making, including tempering clay with other materials such as sand or bone dust, increasing firing temperatures, and glazing pots all increased their range of uses, and by the middle of the Neolithic, ceramic pots were present in many functionally specialized shapes such as jugs, lamps, cups, platters, and a host of decorative, artistic, and representational objects (Violatti 2014). By transforming food processing and storage, and cultural life, ceramics served as profound agents of system change.

4.4.2 *Domestication*

Successful agriculture requires the long-term planned manipulation of DNA. Changing the genetic makeup of a useful array of plant and animal species is a complex undertaking. First you have to find things that are suitable for domestication (Zeder 2006). Any chosen species must have some combination of traits that can be enhanced by selective breeding and physical manipulation to meet a human need (Zeder 2012). All vascular plants have some combination of flowers, stems, fruits, roots, leaves, and seeds. Yet only a few are non-toxic, digestible by humans, and sufficiently nourishing to be worth the effort of cultivation (Abbo et al. 2014). To be useful as domesticates plants must have "larger edible parts, easier harvesting, and decreased defenses" (Stetter et al. 2017: 896). All mammals produce milk and can bear burdens. Only a few can be convinced to lactate continuously and will allow humans to milk them. Only a few willingly bear the weight of a human on their backs. All birds lay eggs. Some eggs are much harder to retrieve than others, for various reasons. Diamond summarizes the problem thus (Fig. 4.2):

> The six main obstacles proved to be a diet not easily supplied by humans (hence no domestic anteaters), slow growth rate and long birth spacing (for example, elephants and gorillas), nasty disposition (grizzly bears and rhinoceroses), reluctance to breed in captivity (pandas and cheetahs), lack of follow-the-leader dominance hierarchies (bighorn sheep and antelope), and tendency to panic in enclosures or when faced with predators (gazelles and deer, except reindeer). Many species passed five of these six tests but were still not domesticated, because they failed a sixth test. Conclusions about non-domesticability from the fact of non-domestication are not circular, because these six obstacles can be assessed independently. (Diamond 2002: 702).

Suitable candidate species of plants and animals must then be carefully bred over many generations to enhance desirable traits, and to diminish undesirable traits. At the same time potential competitors, predators, and parasites must be managed. During this lengthy period of genetic engineering the target species must be successfully woven into a landscape economy that maintains a sustainable ecosystem while providing for the ongoing nutritional needs of the aspiring farmers. This must be accomplished in a way that is also socially sustainable. This is not a quick process. Rather, as Graeber and Wengrow point out, it can be an episodic, reversable, and haphazard process (Graeber and Wengrow 2021).

Favorable Characteristics

1. Social Structure
   a) Large gregarious social groups
   b) Hierarchical group structure
   c) Males affiliated with social group
2. Sexual Behavior
   a) Promiscuous mating system
   b) Males dominant over females
   c) Sexual signals provided by movement or posture
3. Parent-Young Interactions
   a) Social bonds created through imprinting
   b) Female accepts young soon after parturition or hatching
   c) Precocial young
4. Feeding Behavior & Habitat Choice
   a) Generalist feeder or omnivorous
   b) Wide environmental tolerance
   c) Non-shelter-seeking
5. Responses to Humans
   a) Short flight distance from humans
   b) Low reactivity to humans or sudden changes in environment
   c) May solicit attention
   d) Readily habituated

Unfavorable Characteristics

1. Social Structure
   a) Family groupings
   b) Territorial structure
   c) Males in separate groups
2. Sexual Behavior
   a) Monogamous mating system
   b) Females dominate males/males appease females
   c) Sexual signals provided by markings or morphology
3. Parent-Young Interactions
   a) Social bonds created on basis of species characteristics
   b) Female accepts young on basis of species characteristics
   c) Altricial young
4. Feeding Behavior & Habitat Choice
   a) Specialized dietary preferences or requirements
   b) Narrow environmental tolerance
   c) Shelter-seeking
5. Responses to Humans
   a) Extreme wariness and long flight distance
   b) Easily disturbed by humans or sudden changes in environment
   c) Independent/avoids attention
   d) Difficult to habituate

Fig. 4.2 Characteristics of Animal Candidates for Domestication (Zeder 2012)

There are nesting levels of change visible through domestication systems at several scales. The domestication of yeast stands as one of the less researched microbiotic manipulations (Liti et al. 2009). There is a marked change in the microbiome of the human body associated with this change, as human gut flora adapted to the new regime (Mancabelli et al. 2017). One step up from this the cloud of parasites, symbiotes, and commensualists that surround humans was transformed by urbanism. For example, the common house mouse was unintentionally "domesticated" (Weissbrod et al. 2017). This spectrum of change propagates outwards from known examples to the rest of the ecosystem (Dekel et al. 2017).

Leaving aside the complex question of dogs, the economic experiment of domestication was probably first successful in the Zagros Mountains of Iran. Three things helped push the change forwards. First, there was a rich enough ecosystem to support sedentary hunter-gatherers: once settled people in these communities had the time and the conditions in which to pursue the genetic experiments required to develop agriculture. Secondly, there were at least two species suitable for domestication in the general area, einkorn wheat and the mouflon sheep, providing the original stock from which modern domesticated wheat and sheep were bred (Riehl et al. 2013). Thirdly, there was an increase in population density in the region over time, increasing pressure on resources, so driving people to seek new methods of meeting food needs.

Following this general model agriculture developed independently in several other parts of the world. In the Americas this first started around 10 Kya, subsequently developing in at least three regions: South America, Mesoamerica, and eastern North America. There are some key differences in the triggering and sequencing of early American agriculture. The first agriculturalists in America became sedentary only after they had domesticated their first crops, in this case squash, and later maize (Price 2009). Lacking the variety of wild animals suitable for domestication available to their Eurasian and African counterparts, cultivation in the Americas was based on, and remained dominated by, human labor, rather than the use of draft animals. However, the fact that the work was undertaken during the same time frame is notable. Some triggering combination of environmental and social change made the shift to agriculture a viable path forwards for widely separated and isolated peoples at about the same time.

There was a positive feedback loop between agriculture, sedentism, and population which can be simply stated as more food created the conditions for higher human fertility, resulting in a larger settled population, which created the need for more cultivated land, which produced more food, and so on. In the Fertile Crescent people entered the loop at the point of settlements, exploiting abundant wild foods available to permanent communities, gradually developing other methods of providing for the needs of the growing population. In the Americas they probably entered the loop at the point of cultivation, and then created increasingly permanent settlements to more effectively manage cultivation to support population growth. The transition to full agriculture was lengthier, and in some cases never happened. In addition to domesticated species in many American cases a varying array of wild resources were also used. At one site domesticates were enhanced or supplemented with wild animal protein from fish, bivalves, snails, deer, turkey, raccoon, rabbits, and squirrels, while hickory, walnut, and acorns provided plant protein (Price 2009).

The shift to agriculture occurred on the eastern side of the Atlantic first, possibly because the climatic changes associated with the Northgrippian occurred there first. Agriculture, having first emerged in Iran around 10.7 Kya only really started to spread around 8.2 Kya. It then leapfrogged across Levantine, North African and Southern European landscapes, with residual pockets filling in behind innovators as they developed successful agricultural communities. As noted in the last chapter, the technological transition from exploiting predomesticated forms of wild taxa to domestication occurred between 12 Kya and 9.6 Kya. In higher latitudes this timeframe was characterized by Mesolithic cultures, in the fertile crescent it was characterized by the pre-pottery Neolithic. The spread of Neolithic cultures beyond was driven by innovation, and the diffusion of ideas at least as much as it was driven by migration. It reached Central Anatolia by 10,300 BP (Feldman et al. 2019). Some people actively resisted the development, but other groups abandoned hunting and gathering to adopt the new way of life. Throughout the region genetically distinct populations adopted the new way of life (Baird et al. 2018).

In summary, domestication first appeared in the Fertile Crescent, because of the presence of suitable species, and a settled population. Once the technology of agriculture was obtained it spread outwards from the region of innovation, overwhelming the competing land uses of hunter gatherers. On both an individual and on a societal level it was probably a step downwards. However, on a technological and geographical level it was a system of production that consistently overwhelmed its rivals. Coupled with the evolution of crowd diseases in dense human populations, frequently caught from similarly dense populations of domesticated animals, agriculturalists tended to overwhelm their gatherer-hunter neighbors. The spatial impact of the development of agriculture is unavoidable. Once developed the ecological dominance of agriculture spread outwards from the hearth regions to areas which turned out to be better suited for cultivation.

4.4.3 Villages and Agroecosystems

Sedentism was one necessary accompaniment to the development of viable long-term agriculture. A successful agricultural enterprise simultaneously required the creation of an enduring system-level ecological transformation, or agroecosystem. As with all of the other transformations described here this process had started in a few scattered locations before the Northgrippian. There are five known "urban" sites from the Greenlandian that are close to the NAPT: 'Ain Ghazal, Beidha, and Jericho (all in Jordan), and; Çatalhöyük (Southern Anatolia), and Choirokoitia (Cyprus). At least one other site exists in Baluchistan, Pakistan, but that's beyond the scope of this discussion. Pre or proto-agricultural sedentism, as noted previously, is associated with areas of the world where unusually rich ecosystems provided sufficient resources for people to reduce their foraging range, and stay in one place, pushing population densities up. These are the places where the problem of agriculture vs nutrition was solved. That this solution is nuanced at best, and sometimes proves to be a losing proposition, with agriculture periodically underproducing gathering and hunting has been argued by researchers since at least the 1970s (Sahlins 1972) (Fig. 4.3).

The success or failure of the system-level settlement and agriculture combination is a balancing act, which requires communities to manage the population pressure resulting from early agricultural success eventually against overburdening the landscape. Once this happens it is necessary to develop some means of generating more human-friendly calories per km^2, which in turn drives the development of more intense agriculture. The key geographical change associated with the emergence of permanent agricultural settlements is the division of the landscape into a realm of production and a realm of consumption. Successes are, according to Diamond, attributable to "biogeographic luck" (Diamond 2002). Whether luck was a cause, or some other behavioral or cultural variables exerted an influence, is difficult to

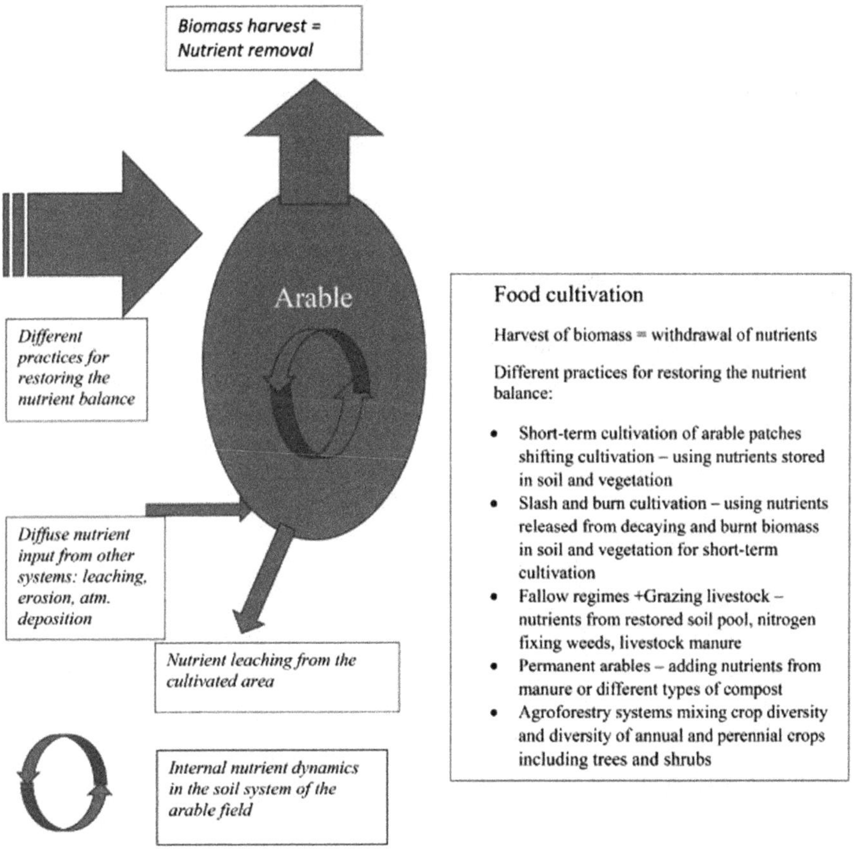

Fig. 4.3 The generalized nutrient challenge in arable agriculture (Olsson 2018)

ascertain at such a historical distance. Regardless, the tendency to develop such arrangements became more marked as people moved into increasingly concentrated, populous, and fortified settlements. However, the trend was established from the outset – there was a place to live and interact with people, and there was a place to work, and interact with nature. The non-urban part of the geographical structure was as essential in the first settlements as it is today. The realm of production was a carefully designed and closely managed agroecosystem.

The visible and deliberate creation of an agroecosystem is only part of the overall change associated with agriculture and settlement. The resulting ecological transformation extends far beyond consciously built domesticated biome, creating a shadow system. In this second system the effects of the changes are visible in such things as exterminated predators, removed competitors, novel parasites, monocrops, and altered hydrology. When successful the agroecosystems evolves as a long-term sustainable structure. However, in some cases it can result in what has been called long-term ecological suicide (Diamond 2002). Diminishing soil fertility, increasing salinity, compromised water supplies, and the disruption of previously productive natural forests and grasslands by cultivation are all part of this shadow system. The successful cultivated landscapes emerging during the Northgrippian can be seen as an unprecedented technological whole, with villages, agroecosystems, and shadow systems all working in concert.

4.4.4 Smelting

The fourth transformational change was metallurgy. The metals first regularly used by people are sometimes called the seven "Metals of Antiquity": gold, copper, silver, tin, lead, iron, and mercury (Mainz 2015). Gold and copper can be found in their native state, so these were the first metals used by humans. As with the other molecular level manipulations discussed here, the first evidence of metallurgy dates back to the beginning of the Greenlandian, with heated and worked copper used to produce ornaments (Craddock 2000). There is no evidence to indicate that this resulted in a widespread transformation of other practices. No tools were made from these materials.

From early in the Northgrippian there are examples of native metals worked, heated, and possibly cast into shapes from a number of locations in the Fertile Crescent, and beyond (Rehren et al. 2013). However, the complex task of smelting, the extraction of metals from ores, requires the breaking of chemical bonds, and then management of the subsequent reactions. It is a multi-step process, requiring controlled high temperatures, the addition of other materials as fluxes, and a structured process of cooling and forming the resulting metal. The first known smelting is evidenced by copper slag, from Serbia (Radivojević et al. 2010). A little later native copper and lead-rich minerals such as galena worked into ornaments show up in Çatal Höyük. Copper and lead are poor substitutes for stone as tools, and so they were first used as decorative objects. A similar pattern of innovation occurred in the

Americas, where first working native metals and subsequently smelting both developed in number of places independently. There is no evidence that the earliest American metalworkers, in the Andes and in Peru influenced the subsequent development of metallurgy in the American NAPT, in Central America. As in Eurasia the first metals were not used for toolmaking, as they did not outperform stone and wood.

Somewhere between 7 Kya and 6 Kya metallurgy underwent a revolution, when copper was successfully smelted from ore. The earliest smelting, as previously noted, occurred in Serbia. The earlies known copper tools were associated with the Vinča culture, centered on the Serbian copper mines. The technology spread to the fertile crescent, and into the rest of Europe, when it became the hallmark of the cluster of cultures collectively described as Chalcolithic. The copper tools of the Chalcolithic were effective, and widely adopted. They are associated with one of the earliest cultures of which we have a complex and nuanced record, in the person of Ötzi, the ice man, the world's oldest known preserved human. Ötzi, among other things, is dressed for travel, and is carrying a full toolkit, including a sophisticated functional copper axe from around 5.3 Kya. He has a copper dagger, and a longbow. He also has an array of well-used stone tools, arrowheads, equipment to maintain them, medicines, fire-making kit, and foul weather gear. Ötzi tells the story of an interconnected culture, with agriculture, pastoralism, and hunting all comprising part of the lifeway. Physiological evidence and artifacts show that Ötzi lived in a web of geographical and social connections that extended across much of the Italian Peninsular (Wierer et al. 2018). Metals existed within this web, but as part of a more complex technological suite, that included many other materials and tools.

In North America the Old Copper Complex culture, centered around the Great Lakes, developed a sophisticated metal-tool technology between 6 Kya and 3 Kya. Uniquely North Americans reverted to the use of stone tools after three millennia. There are several hypotheses as to what drove this shift. As previously noted, copper tools are not necessarily more efficient than their stone equivalents. In the case of the Old Copper Culture it appears that population growth and shifting social dynamics may have created an "accident of history" in which copper production became increasingly focused on the production of ornamental items, at the cost of tools (Bebber et al. 2019). Everywhere else the functional superiority of metal tools and weapons, particularly after the development of bronze, were a driving force in expanding the reach of agricultural societies.

There is a great deal of research exploring the broader implications of developing metal technologies. Among other things smelting impacted the ways in which societies were organized and coincided with what seems to be an increase in social differentiation – concepts of wealth and class were fundamentally changed by the development of metal technology. The presence of valuable metal items in burials and caches indicate that acquisition of valuable tools and ornaments was one way in which some people distinguished themselves from other people (Graeber and Wengrow 2021). The development of complex and long trade networks also seems to have increased significantly at this time. The raw materials for copper smelting, malachite or native copper, are bulky and fairly rare. The people who lived where these things could be found, and who also knew how to work them, were in a position to enhance their skills through practice, and to trade the resulting materials to others. There is an extensive commentary in the archaeological literature as to what these transitions looked like, and how they played out from place to place. Thus, as previously noted in North America the Old Copper Complex culture ended up focusing all production on decorative goods, and finally abandoned metal tools altogether, reverting to stone. In other parts of the NAPT the result was different. Copper smelting birthed an entire social structure based on manufacturing, trading, and using the new tools. In the context of this book this discussion falls outside my area of interest. The development of metals has a marked landscape impact based on their ability to transform the efficacy of human labor. Metal tools increased the productivity of each person. You can cut more wood in less time with a copper axe than with a stone axe. For all the drawbacks of metal tools – they are expensive to make, and they enmesh the user in a complex set of social, economic and geographical networks that many reasonable people might choose to avoid – they have a huge impact on how people interact with their environment. The concentration of individual effort impacted landscapes throughout the Fertile Crescent and beyond. Beyond the more complex cultural change metallurgy was a force multiplier and so became an instrument of landscape change.

The transformation of life and landscape driven by the Copper Age was amplified in Eurasia with the development of bronze, the alloy of copper and tin. Bronze tools were harder to make, requiring higher temperatures for smelting and casting. They took longer to finish and worked better. Being harder they could be sharpened more effectively and held an edge longer. The Bronze Age is notable for being the time during which there is the first evidence of the widespread use of specialized weapons. It is notable that Ötzi's dagger showed no marks of use for cutting plants or other "day-to-day" tasks, as did the rest of his toolkit. It was also carried on his belt, not in his belt-pouch. It has been suggested that this may be because the dagger was some kind of status symbol (Wierer et al. 2018). Given his violent end Ötzi was also a fighter. Weapons were part of his culture also. That is, Bronze Age swords represent the first tool that we know of that was designed and manufactured for the sole purpose of killing other people. The abundance of Bronze Age swords through Europe indicate the development of a specialized trade in fighting. There is good evidence that this trade was a carefully studied and had its innovators and rival schools, as with any other trade (Hermann et al. 2020).

4.5 End of the Northgrippian

The Northgrippian ended with the 4.2 k event, a protracted period of cooling and drying, lasting at least a century. It does not have as clear a signature in the Greenland ice core as the 8.2 k event, however it does show up there, and there is widespread other evidence of its global effects. It also represents the penultimate change in the gradient of the ζ curve. From the 4.2 k event forwards until the twentieth century global average temperatures were significantly cooler than during the

Northgrippian. There is significant (but not universal) agreement that many regions underwent substantial environmental and cultural changes at this time. It has been suggested that Ancient Egyptian Culture, Mesopotamian Culture, and Ancient Indian Culture were sufficiently disrupted to go into a significant decline. Further, there are suggestions that some Chinese Neolithic cultures also suffered a decline at around this time. The event did not precipitate a single simultaneous global catastrophe. It did, however mark a turning point in a number of core cultural regions, seemingly ushering out Neolithic agriculture and settlements, and opening the way for the larger and more complex societies of the Meghalayan (Ran and Chen 2019).

References

Abbo S, van Oss RP, Gopher A, Saranga Y, Ofner I, Peleg Z (2014) Plant domestication versus crop evolution: a conceptual framework for cereals and grain legumes. Trends Plant Sci 19(6):351–360

Anderson DG, Russo M, Sassaman KE (2007) Mid-Holocene cultural dynamics in southeastern North America. In: Climate change and cultural dynamics. Academic, Amsterdam, pp 457–489

Baird D, Fairbairn A, Jenkins E, Martin L, Middleton C, Pearson J, Asouti E, Edwards Y, Kabukcu C, Mustafaoğlu G, Russell N (2018) Agricultural origins on the Anatolian plateau. Proc Natl Acad Sci 115(14):E3077–E3086

Bebber MR, Key AJ, Fisch M, Meindl RS, Eren MI (2019) The exceptional abandonment of metal tools by north American hunter-gatherers, 3000 BP. Sci Rep 9(1):1–4

Büdel J (1973) Climatic geomorphology. Palgrave, London

Craddock PT (2000) From hearth to furnace: evidences for the earliest metal smelting technologies in the eastern Mediterranean. Paléorient 2000:151–165

Dekel Y, Machluf Y, Brand R, Partouche ON, Ben-Shlomo I, Bercovich D (2017) Mammal domestication and the symbiotic spectrum. Proc Natl Acad Sci 114(27):E5280

Diamond J (2002) Evolution, consequences and future of plant and animal domestication. Nature 418(6898):700–707

Feldman M, Fernández-Domínguez E, Reynolds L, Baird D, Pearson J, Hershkovitz I, May H, Goring-Morris N, Benz M, Gresky J, Bianco RA (2019) Late Pleistocene human genome suggests a local origin for the first farmers of Central Anatolia. Nat Commun 10(1):1–10

Graeber D, Wengrow D (2021) The dawn of everything: a new history of humanity. Penguin, London

Harari YH (2018) Sapiens: a brief history of humankind. McLelland and Stewart

Hermann R, Dolfini A, Crellin RJ, Wang Q, Uckelmann M (2020) Bronze age swordsmanship: new insights from experiments and Wear analysis. J Archaeol Method Theory 17:1–44

Kalis AJ, Merkt J, Wunderlich J (2003) Environmental changes during the Holocene climatic optimum in central Europe-human impact and natural causes. Quat Sci Rev 22(1):33–79

Klein K, Goldewijk K, Beusen A, Van Drecht G, De Vos M (2011) The HYDE 3.1 spatially explicit database of human-induced global land-use change over the past 12,000 years. Glob Ecol Biogeogr 20(1):73–86

Liti G, Carter DM, Moses AM, Warringer J, Parts L, James SA, Davey RP, Roberts IN, Burt A, Koufopanou V, Tsai IJ (2009) Population genomics of domestic and wild yeasts. Nature 458(7236):337–341

Mainz VV (2015) The metals of antiquity and their alloys. In: Chemical technology in antiquity. American Chemical Society, pp 139–179

Mancabelli L, Milani C, Lugli GA, Turroni F, Ferrario C, van Sinderen D, Ventura M (2017) Meta-analysis of the human gut microbiome from urbanized and pre-agricultural populations. Environ Microbiol 19(4):1379–1390

Marcott SA, Shakun JD, Clark PU, Mix AC (2013) A reconstruction of regional and global temperature for the past 11,300 years. Science 339(6124):1198–1201
Medvedev VE, Tsetlin YB (2013) Technological analysis of the earliest ceramics from the Amur region (13–10 thousand years BP). Archaeol Ethnol Anthropol Eurasia 41(2):94–107
Mendieta KL, Gerber S, Brenner M (2018) Florida wildfires during the Holocene climatic optimum (9000–5000 cal yr BP). J Paleolimnol 60(1):51–66
Olsson EG (2018) The shaping of food landscapes from the Neolithic to industrial period: changing agro-ecosystems between three agrarian revolutions. In: Zeunert J, Waterman T (eds) Routledge handbook of landscape and food. Routledge, London, pp 24–40
Pielou EC (1991) After the ice age. University of Chicago Press, Chicago
Price TD (2009) Ancient farming in eastern North America. Proc Natl Acad Sci 106(16):6427–6428
Radivojević M, Rehren T, Pernicka E, Šljivar D, Brauns M, Borić D (2010) On the origins of extractive metallurgy: new evidence from Europe. J Archaeol Sci 37(11):2775–2787
Ran M, Chen L (2019) The 4.2 ka BP climatic event and its cultural responses. Quat Int 521:158–167
Rehren TH, Birch T, Pernicka E (2013) The metallic finds from Çatalhöyük: a review and preliminary new work. In: Hodder I (ed) No B M substantive technologies at Catalhöyük. British institute at Ankara, Monograph no. 48 Monumenta Archaeologica 31. Cotsen Institute of Archaeology Press, Ankara, pp 307–316
Riehl S, Zeidi M, Conard NJ (2013) Emergence of agriculture in the foothills of the Zagros Mountains of Iran. Science 341(6141):65–67
Sahlins MD (1972) Stone age economics. Aldine, Chicago
Schaney ML, Kite JS, Schaney CR, Thompson JA (2021) Evidence of mid-Holocene (Northgrippian age) dry climate recorded in organic soil profiles in the central Appalachian Mountains of the eastern United States. Geosciences 11(11):477
Stetter MG, Gates DJ, Mei W, Ross-Ibarra J (2017) How to make a domesticate. Curr Biol 27(17):R896–R900
Tricart J, Cailleux A (1972) Introduction to climatic geomorphology. Longman, London
Violatti C (2014) Pottery in Antiquity. World History Encyclopedia. https://www.worldhistory.org/pottery/. Accessed 16 May 2022
Walker M, Head MJ, Lowe J, Berkelhammer M, Björck S, Cheng H, Cwynar LC, Fisher D, Gkinis V, Long A, Newnham R (2019) Subdividing the Holocene series/epoch: formalization of stages/ages and subseries/subepochs, and designation of GSSPs and auxiliary stratotypes. J Quat Sci 34(3):173–186
Weissbrod L, Marshall FB, Valla FR, Khalaily H, Bar-Oz G, Auffray JC, Vigne JD, Cucchi T (2017) Origins of house mice in ecological niches created by settled hunter-gatherers in the Levant 15,000 y ago. Proc Natl Acad Sci 114(16):4099–4104
Wierer U, Arrighi S, Bertola S, Kaufmann G, Baumgarten B, Pedrotti A, Pernter P, Pelegrin J (2018) The Iceman's lithic toolkit: raw material, technology, typology and use. PLoS One 13(6):e0198292
Zeder MA (2006) Central questions in the domestication of plants and animals. Evol Anthropol 15(3):105–117
Zeder MA (2012) The domestication of animals. J Anthropol Res 68(2):161–190

Chapter 5
The Meghalayan

Historical Time (4200 BP – Present)

Abstract The final age of the Holocene. The natural forcing of climate episodes, including the Neoglacial, the Roman Warm Period, the Dark Ages Cold Period, the Little Ice Age. The humans forcing of subsequent climate and ecosystem change. The Industrial Revolution. The establishment of widespread agroecosystems throughout the NAPT that are now triggering system-level changes in atmosphere, hydrosphere, cryosphere, and biosphere.

Keywords Neoglacial · Roman warm period · Dark ages cold period · Little ice age · Industrial revolution

5.1 Introduction

The Meghalayan is the third and current stage of the Holocene, and has so far lasted 4200 years. The boundary between the Meghalayan and its predecessor, the Northgrippian is defined by the Global Boundary Stratotype Section and Point (GSSP) at the Krem Mawmluh Cave formation in Meghalaya, northeast India. The defining characteristic of the Meghalayan is the falling arm of the ζ-curve, and the current abrupt upward spike (Fig. 3.1). Throughout most of the stage, there is a decline in temperature of around 0.5° globally. As with all such broad-brush descriptions this general pattern contains a host of small variations, several of which show up in the NAPT, and are of historical significance – for example the Medieval Warm Period, and the Little Ice Age. Some of these fluctuations have left a clear geological trace. Others are better known from the human record. This is because the Meghalayan contains almost all recorded human history – that is, the written records of human actions. There are areas of the world such as the Americas, and parts of the Pacific where writing systems were developed, but no-one knows how to read them any longer. There are also areas of the world where it seems writing was never developed. This is sometimes the case where agriculture was not widespread. However, all of the complex civilizations of the classical world, and many others developed written scripts. The Meghalayan marks the widespread move from

M. Bampton, *The North Atlantic Polar Triangle*, Springer Polar Sciences,
https://doi.org/10.1007/978-3-031-27264-6_5

village agriculture to full-scale hierarchical feudal structures. It also marks the time during which nearly every habitable ecosystem capable of sustaining human life was populated. In environmental terms the human thumbprint extended from the transmutations of the Northgrippian and evolved into the conscious and planned reconstructions of entire ecosystems. By any measure the Meghalayan is the age of human acceleration within the NAPT.

In the Meghalayan there was, by any measure, an astonishing arc of human experience. At its beginning the earliest complex agricultural civilizations were only just established. By its end modern industrial society dominated all but a very few places on dry land, and had profoundly altered the oceans and the atmosphere. It is the age in which humans have finally, and definitively, touched every part of the Earth's surface. It is also the age in which humans have started to extend their reach beyond the planet. In the Meghalayan humans emerged as the most pervasive and persistent forcing agent in an extraordinarily broad array of natural processes. Among other things, the end of the Meghalayan is a human artifact. To manage such a rapid increase in complexity this chapter requires some additional organizing structures not used in the two that preceded it. As well as arranging things topically, I shall organize them by time. Individual topical characteristics of the Meghalayan can also be divided into several time slices, that are defined by a combination of physical and cultural markers. Thus, for instance, there are a number of warm periods, such as the Roman Warm Period, and the Medieval Warm Period, that coincide (more or less) with clearly identifiable periods of human history. To make things clearer I shall adjust the dating convention I have used until now, and add calendar dates "before present (BP) to the conventional abbreviation of Kya for thousand years ago. Many events are more easily recognizable in their historical context using this method of dating.

Although these are only infrequently visible across the entire region, there are some general patterns that emerge. I am not a historian, and this is not a history book, at least not in any conventional sense, so these mismatches are less important here than they might at first seem. Despite some differences in the timing and sequencing of events, from a natural historical perspective there are several broad trends that characterize the NAPT during this time:

- The ascendency of human action in natural processes
- The demographic dominance of humans
- The prevalence first of agriculture, then of industry as modes of production
- The increasing speed and frequency of human interactions over the Earth's surface
- The increasing speed and frequency of the human-managed movement of inorganic and organic materials over the Earth's surface
- An increasing simplification of terrestrial ecosystems
- The genetic ascendancy of domesticated and cultivated species over their wild counterparts, competitors, and progenitors.

The dynamics of change during the Meghalayan increasingly tip towards human causation. Where exogenous forcing agents play a role, they are limited in scope and effect. Climate fluctuations, for example, driven by astronomical or volcanic processes have decreasing impacts in terms of broad trajectories of processes. The general trend towards expanding human population remains on track. The trend of increasing complexity and prevalence of agriculture remains set. When individual societies change, or collapse, they are frequently replaced by analogous complex societies within a few generations. Even the most catastrophic failures are temporary, and the prevailing population dynamics of the Meghalayan have remained consistent. Human population has grown exponentially for over four thousand years.

5.2 Climate

The general character of Meghalayan climate is summarized in the falling arm of the ζ-curve. That is, there has been a prevailing gradual cooling trend for most of the past 4000 years or so. One of my central arguments in this book is that the terminal spike of the curve, when considered in combination with numerous other inflection points on other curves recording other anthropogenic changes, represents the emerging upper chronostratigraphic boundary of the Meghalayan, and probably also of the Holocene. Figure 5.1 shows, overall the granularity of the record changes about halfway through the stage. This is because in general, the most recent 2000 years of the Meghalayan are understood in much greater detail, and the Northern Hemisphere is much better studied. The primary reason for this change in the character of the data boils down to the presence of significantly better records of all kinds. Sources including living trees, lake and ice cores, pollen, faunal remains, and human historical records are all present for earlier in the Holocene, but they are much more abundant for this stage. This can make comparisons difficult, as the first 2000 years of the stage are not commensurate with the second 2000 years.

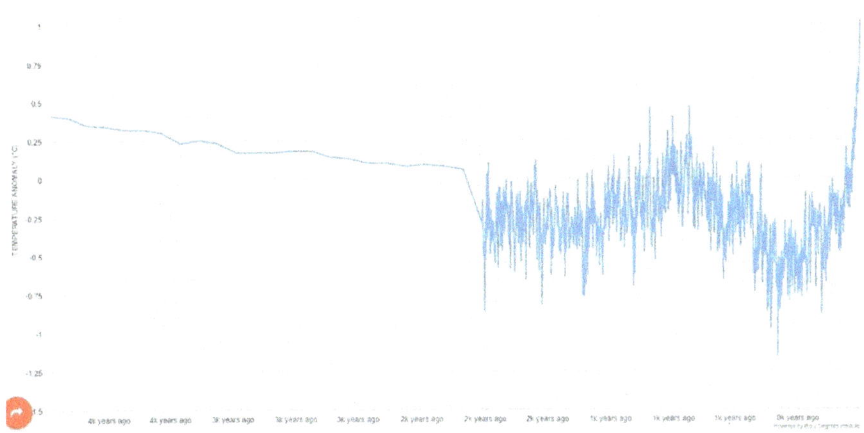

Fig. 5.1 Global average temperature in the Meghalayan (After https://www.temperaturerecord.org/)

Peaks, troughs, trends, "warm", and "cold" periods can all be defined more easily. During the most recent two millennia the increasing frequency of multi-year climate fluctuations is almost certainly an artifact of the improved record. Documented Meghalayan climate anomalies attributable to process changes in the NAPT start with the general cooling trend or neoglacial following the 4.2 k event. As the record becomes more detailed five other named periods are identifiable. All reference European history in their names, as these definitions rely to some extent on written records which are far less common or completely absent in antiquity for both Africa and the Americas. This Eurocentric tilt is slowly being rectified in the record, as physical proxies are brought to bear in confirming the record. The region five historical periods as described by Spielhagen et al. are (Spielhagen et al. 2011):

1. 2200 BP Roman Warm Period (RWP 250 BC ~ 600 CE)
2. 1350 BP Dark Ages Cold Period (DACP, ~600 to ~900 CE)
3. 1050 BP Medieval Climate Anomaly (MCA, ~900 to ~1500 CE)
4. 450 BP Little Ice Age (LIA, ~1500 to ~1900 CE)
5. Modern (Industrial) Period.

Overall the Meghalayan range in global temperature averages ran from a little under 1° colder than the Holocene average, to around 0.5° warmer. It also saw the end of the African Humid Period (AHP ~ 14,000 BP to 4000 BP), which lasted from the LGM through the Greenlandian and the Northgrippian. In South America (notably in the Atacama region of Peru, outside the NAPT) the Late Holocene Wet Phase roughly matched the Meghalayan (Sáez et al. 2016). All of these shifts were almost certainly driven by combination of Milankovitch forcing, and internal system triggers such as volcanic activity, and NAO/AMOC patterns. At this level of granularity process lags in time and space, and the relative importance of different factors become increasingly apparent, and local expressions of complex trends become more common. This can confuse the interpretation of the record somewhat. On the one hand it is possible to change the granularity of analysis to match the character of the data. The literature is rich with analyses of the impacts of smaller climate changes on smaller areas over shorter times. In many instances very small shifts can have catastrophic consequences for local marginal systems. That is, tipping points can be reached in both human and natural systems with even small shifts in climate parameters (Bampton et al. 2018). The chain of causation from these localized changes can be traced back up towards shifts in the behavior of regional systems. This is significant, as it is therefore reasonable to argue that, even if not evident in the record at present similar "trickle down" catastrophes occurred earlier in the Meghalayan, and before. The general pattern can reasonably be stated that system-wide natural changes change can drive small local crises. When these accumulate within a broader directional cultural trend can have systemic outcomes.

The most striking trend of the Meghalayan climate is the ascendancy of humans as a forcing factor. There is a lively debate about when human intervention became a significant agent of climate change. It is possible to reason from first principles that humans have been influencing climate at least on a local level from the advent of agriculture onwards. Clearing land for cultivation changes albedo – the radiation reflection characteristics of surfaces – and immediately influences local energy dynamics. Likewise, changing vegetation by grazing, planting, harvesting, or by diverting water to irrigate crops changes local patterns of evaporation and transpiration. It also alters the atmospheric carbon budget. The point at which these shifts in energy and water budgets become sufficient to alter regional or global climate is debatable. There is some recent research indicating that the influence of preindustrial agriculture on climates may be greater than previously believed. Whatever the outcomes of this debate, the unavoidable (and inconvenient) truth is that by the end of the Meghalayan humans have, for the time being, become the primary climate forcing factor.

To unpick these dynamics, it is useful to explore each of these Meghalayan time-slices as they are most evident within the NAPT. This will provide some useful context for the changes that occur at the end of the Meghalayan, as there are some distinct differences in the scale, timing, and extent of each of these changes. All five of these changes, though trackable throughout the NAPT do not have common time signatures across the rest of the globe, and the currently available evidence suggests they overlap in time and space. They are, then, examples of events that have a local signature but do not rise to the level of chronostratigraphic boundaries.

5.2.1 The Neoglacial

Following the 4.2 k event, and lasting until about 2000 BP there was a general trend of glacial advance in the Arctic. Recent compilations of a number of studies give an indication of the complexity of non-boundary events such as the neoglacial. While having widespread regional effects in the high latitude Northern hemisphere it did not rise to the level of a chronostratigraphic boundary. This is sometimes described as a neoglacial, as there was a marked advance of ice sheets in high latitude parts of the continents on both sides of the Atlantic. However, at this resolution, the complexity of climate change becomes very clear. It was not a single synchronous effect throughout the region. But there is good physical evidence to indicate that the global cooling trend that started before (and therefore perhaps defined) the 4.2 k event around 5000 BP, continuing until at least 2000 BP (Bradley and Bakke 2019), had a strong regional signature (McKay et al. 2018). The glacial advance was not uniform in time or space, and its effect were limited to higher latitudes. This suggests that a complex series of mechanisms were at work. This probably included a decline in insolation, and increase in reflective ice cover, and the broader regional climate dynamics of the time. That is, shifts elsewhere within this linked system resulted in a regional non-synchronous cooling event.

5.2.2 The Roman Warm Period

There was a period of unusually warm weather in the eastern Mediterranean and Aegean seas over a period of around 850 years (Ljungqvist 2010). Starting about 2200 BP summer temperatures in much of the core region of the expanding Roman Empire were about 1° warmer the Holocene average. There are some suggestions from historical sources that it may also have been a more humid time, with rainfall increasing throughout much of the region. However, this was not a simple matter of increased temperatures over the whole NAPT, or even over Roman Europe. Rather, it was a case of there being some marked trends towards warmth and humidity. It was also a time of significant transformation of the other landscape parameters, as the success of Roman agriculture and the efficiency of the taxation system meant that agricultural yields were significantly increased throughout the Empire. Evidence of this can be seen in increased rates of soil erosion, from the Pyrenees to the Roman "breadbasket" of North Africa. There is some evidence of comparable warming in North America in roughly the same time period. However, the prevailing impression is that the RWP was not synchronous throughout the NAPT.

5.2.3 The Dark Ages Cold Period

Dark Ages Cold Period (DACP) followed the Roman Warm Period, and ran from around ~600 to ~900 CE, or from 1350 BP – 1050 BP. There is credible evidence to indicate that the cooling affected Europe, the North Atlantic and Arctic regions, North America, China/Tibetan Plateau and the northern Pacific. There was a period of contemporaneous wetness in North Africa, and parts of Europe (Gasse 2000). There is a significant body of evidence from a wide variety of locations indicate that there was a notable shift in climate patterns (Helama et al. 2017). The prevailing character of coldness and increased wetness is concentrated in and around the North Atlantic. Though as is so often the case, this may simply be an artifact of available research findings. There is no consensus on how much colder, or how much wetter the DACP was, even in general terms. However, there is clear agreement that there were widespread glacial advances in higher latitudes, an significant shifts in precipitation in lower latitudes. In some cases, conditions were drier, but the general trend was towards increased humidity and precipitation. The candidates for primary forcing agents are El Niño, La Niña, with some influence from volcanic aerosols.

5.2.4 The Medieval Climate Optimum

The medieval climate optimum is sometimes also known as the "medieval warm period". It almost certainly affected the majority of the globe, and was the most recent period in which temperatures came close to matching the current highs.

Consequently, it has become a bone of contention in some circles. It has been argued that the consistent warming trend during the Middle Ages is comparable to modern climate trends, and that therefore contemporary climate change is an example of an ongoing cycle, between warm and cold periods, rather than an anomaly. Broeker suggested that one interpretation of the data yields a consistent 1500 years fluctuation around a mean value throughout the Holocene (Broecker 2001). He speculated that this process may be driven by changes in the thermohaline "conveyor belt" moving water and thermal energy from the tropics to the poles. Within this model the current spate of global warming is predictable, and in keeping with earlier fluctuations. It is, therefore, worth taking some time to explore the MCO in a little more detail, and to consider some of the substantial arguments that have been mounted against it. Available reconstructions and models indicate that the MCO had temperatures somewhere between −0.4° and + 0.4° of the 1961–1991 average, a time slice that has been used as a reference temperature against which to measure its magnitude (Mann et al. 2008). The evidence to support this is much denser for the Northern Hemisphere, and much more complete for the period from 1000 AD forwards. The regional bias of proxy climate data is sufficiently marked that there is a real imbalance in our understanding of how these multi-decadal trends played out over large parts of the world. Figure 5.2 shows a variety of different versions of the Northern Hemisphere proxy records. Two things are immediately apparent. Firstly, the claim that the amount of MCO warming equaled current warming trends is only sustainable in some locations and for some proxies. Lake sediments consistently give a higher temperature estimate for the MCO. Second, MCO temperature values for Siberia and China, among others show comparable increases in temperature. A signal event frequently cited as evidence for the prevalence of the MCO is the settlement first of Iceland, and secondly of Southern Greenland, by migrating Norse from

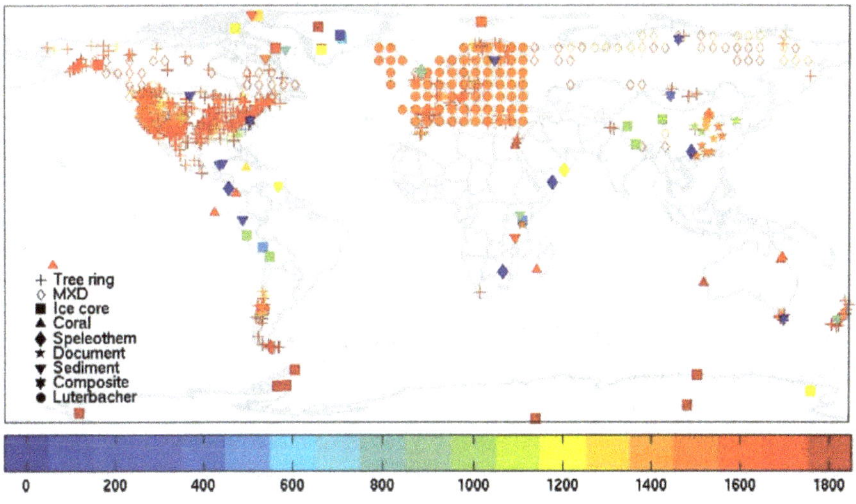

Fig. 5.2 Spatial distribution of Holocene climate proxies (Mann et al. 2008)

Scandinavia. The Norse in both cases brought with them a European survival strategy based on agriculture and pastoralism.

5.2.5 The Little Ice Age

The Little Ice Age (LIA) was a global decline in temperature that occurred between the 1300s and the 1850s. Of all the Meghalayan climate fluctuations this is the best understood. There are many types of records describing its duration and character in many locations within and beyond the NAPT. Its later years fall comfortably within the instrumental period, on both sides of the Atlantic. Even its earliest stages are recorded in written observations and oral accounts. In addition to these human first-hand records there are also excellent proxy records, from tree-rings to pollen, from ice-cores, to lake sediments. Artistic renderings, creative literature, geomorphological evidence, and historical records all combine to create a detailed impression of this period of time.

As with all previous Meghalayan climate fluctuations (and the less well-documented ones from still earlier in the Holocene) it was unevenly distributed in both space and time. However, the general characteristics were the same through the NAPT. Temperatures were on average around 0.5° cooler than at present. This was sufficient to trigger ice advances in higher latitudes and altitudes. It also caused, variously, cooler and wetter conditions in parts of the mid latitudes, and droughts and severe weather elsewhere.

Perhaps most striking about the LIA is its coincidence with the period of European mercantile and colonial expansion. It starts around the time of European maritime exploration outwards from the Mediterranean heartland. It ends around the time when the Industrial Revolution becomes the predominant economic force in the Atlantic region. This coincidence has led to some speculation about the linkages between the two sets of processes, physical and social. However, it is only recently that any credible claim of causal mechanism has been proposed. It has been suggested that there may be some causal links between the LIA and this historical trend. Certainly, the argument can be made that the end of the LIA may have been precipitated by human action. However, arguing that its start or its severity can be attributed to human action remains to be definitively demonstrated. Nonetheless, the argument supporting this notion is intriguing. Koch et al. have presented the case that massive reforestation, following the "Great Dying" (Koch et al. 2019). They argue that there is a causal link between the population collapse in the Americas following European contact, the consequent abandonment of agricultural land, and subsequent reforestation. The massively increased tree coverage of the Americas caused a huge drop in atmospheric carbon, as the vegetation absorbed CO_2. The change in atmospheric chemistry either triggered, or more probably exacerbated, a cooling trend. Only with the return of carbon to the atmosphere following colonial deforestation and increased industrial combustion was sufficient to reverse this cooling trend, and ultimately initiate warming.

As noted with regard to the MCO, one of the signal events is the abandonment of Norse settlements in Southern Greenland. Recent work has revealed a lot about the marginal nature of the Viking economy in Greenland. Most telling is that by the end of the MCO there were few options for a European agricultural society to persevere in the marginal landscapes of Sothern Greenland (Jackson et al. 2018). The extreme case of Norse Greenland is illustrative of a much broader trend. The LIA was characterized by increased storminess and generally colder conditions throughout much of the higher latitude NAPT. The pattern of cooling, and the impacts it had on environmental processes and societies was not consistent for the entire half-millennium. As it is so (comparatively) well documented it is possible to construct a much fuller picture of Rather, it was a protracted and qualitative shift in key process regimes, that fundamentally changed the way in which the world operated.

5.3 Ecosystems

Climate shifts in the Meghalayan translated into ecosystem transformation. One of the most staring examples of this is the aridification of the Sahara. As previously noted, for the bulk of the Greenlandian and the Northgrippian the Sahara was significantly more humid than at present. From around 10,000 BP until 6000 BP it was grassland. However, following the 4.2 K event the region rapidly dried out. The shift from a wet Sahara to a dry Sahara has happened a number of times in the Quaternary. The end of the LGM marked the culmination of a dry period caused by the binding of water in ice sheets. The following lengthy humid period finally ended with the Meghalayan. It is also in the Meghalayan also that the trend of humans as the dominant influence on environmental processes becomes firmly established. This dominance is expressed first and foremost in the widespread transformation of ecosystems. The construction of agroecosystems throughout the lands of the NAPT required a consistent strategy of ecosystem manipulation: the elimination of competing species to create space for cultivation, the enhancement of soils and grasslands to support crops and grazing animals, and the management of hydrological systems to provide for irrigation. These strategies were all well-established – in some cases since the Greenlandian. What was different in this instance was the extent of the transformation. It is worth looking at each in turn.

5.3.1 Eliminating the Competition

A recurring and prevailing myth describes the human struggle against predators. In some cases, this shows up in epic legends, where heroes slay lions and tigers and bears. It also permeates our childhood consciousness. Little Red Riding Hood, Beowulf, Goldilocks, and the heroes of Africa and America all successfully triumph over fearsome wild creatures that are apex predators. The prevalence of this storyline gives a clear indication of how important this struggle was. What distinguishes the Meghalayan was that during this time the battle was decisively won by humans. Rival predators in many ecosystems were marginalized or driven into extinction. The classical hero wrapped in a lion skin, the leopard-skin shoes of a prime minister, the bearskin hat of a soldier, or the jaguar throne all tell the same story of a slain predator. This triumph over equals in the food-chain, if not in the shifting sands of the evolutionary journey, marks the global ascendancy of human society. As noted in earlier chapters humans seem to have been potent forces of extinction from the time they migrated out of Africa. Although other forces may have contributed to the wave of extinctions that coincided with human arrival in new ecosystems there is a growing body of evidence to indicate that the primary forcing factor was human action.

The extent to which humans managed to control vegetation is harder to ascertain, however there are several dynamics that can be identified. The most immediate and lasting impact was forest clearance. With the advent of agriculture there is also, almost always, widespread deforestation. This is firstly to clear land for cultivation. Secondly, also to provide fuel and in those cultures to fire furnaces and kilns for metallurgy and firing pottery. There are several other ways in which vegetational systems are modified by human actions, all of which we have encountered in previous chapters. Genetic manipulation of cultigens has an impact on wild species in several ways. First, there are many species which will still cross-pollinate between wild and cultivated varieties of the same plant. This same crossover potential is true in some cases for parasites, pests, and pathogens. Cultigens are also frequently introduced into environments far from the original source regions of their wild ancestors. As invasive species they can on occasion successfully compete with local species and overwhelm them. This is the case for both intentional and unintentional introductions. Finally, competing wild species are eliminated wherever possible.

5.3.2 *The Enhancement of Soils*

Leaving aside irrigation for a moment, which will be addressed in the next section, the manipulation of soils centers on two factors: the transformation of chemistry, and the transformation of structure. In the first instance, the transformation of chemistry is accomplished by either adding organic or inorganic materials. These can include manure, dead vegetation such as leaf-mold or seaweed, shells, or animal and fish carcasses, and some cropping practices, such introducing nitrogen-fixing species like lupins and beans. Inorganic materials can include calcium, sand, or easily weathered materials such as volcanic ash. On occasion both organic and inorganic materials are naturally added – shells for example in the case of Machair, sand in the case of dune migration, and volcanic ash in the case of eruption. The structural transformation of soils is accomplished by digging, ploughing, letting fields lie fallow, burning or introducing animals and plants that have survival strategies that disrupt or transform soil profiles.

The two strategies in practice almost always work in tandem. A consistent characteristic of agricultural societies of any kind is a long-term investment of labor, typically multi-generational, in this manipulation of soils. Even within the geographical constraints of the NAPT there is an impressive number of variants on this practice. As the number of cultigens increased, and as the range of climatic and ecological conditions in which agriculture was pursued was extended out from the core regions of innovation the extent to which soils were manipulated increased, and the technologies employed in manipulating them became more complex. For most of the Meghalayan the practice of manipulation was based on tradition, observation, and trial and error over long periods of time. The absence of wide-ranging general knowledge underlying the processes of soil engineering is indicated by some of the spectacular catastrophes that occurred in cases where conditions changed. Thus, repeated irrigation in arid environments leading to the upward transport and increasing concentrations of salt at the surface was a common cause of land abandonment in classical farmlands. Misunderstanding of the relationships between deforestation and soil erosion was similarly problematic. Overgrazing, overpopulation, overirrigation, and over manuring all caused difficulties. The fact that the complex hierarchical societies of classical empires almost invariably failed catastrophically indicates the inherent instability of the model of constant growth that is inherent in their structure.

5.3.3 The Management of Grazing Systems

Pastoralism, a specialized form of agriculture, runs the gamut from the management of genetically wild animals such as reindeer through to the rearing of fully domesticated animals. In all cases the pasture which animals survive is an anthropogenic environment. One of the great differences between the agricultural histories of the eastern and western sides of the NAPT is the absence of domesticated pastoralism in pre-Columbian North America. With the exception of the brief Norse sojourn in Greenland the only form of pastoralism practiced in North America was the pursuit of migratory herds of Arctic deer, and prairie buffalo. Although pastoral lifeways did develop in the Andes, this practice did not spread through central and North America until after the arrival of Europeans. In contrast, on the Eastern side of the NAPT there was a wide range of pastoralism practiced using the full gamut of techniques, from transhumance through to herd-following.

There is an important distinction which must also be made when considering pastoralism is between pastoralism as the predominant survival strategy, and pastoralism in conjunction with other strategies. Arctic peoples such as the Sami and the Nenets maintain, harvest, and protect reindeer, still an undomesticated species. In something resembling a symbiotic relationship that persists into the twenty-first century these peoples are guardians of wild animals, harnessing them, culling them, and protecting them. They bridge the gap between pure hunting cultures such as the Caribou Inuit and nomadic herders. Within the NAPT true nomadic pastoralism is quite rare. However, it does exist in a hybrid form paired with other types of agriculture. Thus, in many of the mountainous and high latitude regions of the Eastern NAPT it forms a long-established and effective strategy. As such it is associated with careful long-term land management. Once opened high altitude pastures in particular demand maintenance. The practice of transhumance, moving animals from summer to winter pastures is practiced through much of Europe. In the British Isles it persists in areas with extensive moorlands, including Wales, the Highlands and Islands, Wessex, and Cumbria. It is also common in locations as diverse as Macedonia, the Alps, the Pyrenees, and Scandinavia. It was a major economic force in post-Moorish and pre-American Spain (Butzer 1988). It is also typically described as part of the Norse "survival package", and so shows up in the surviving and the failed Norse settlements of Iceland, the Faroes, and Greenland.

The manufactured landscapes result from these herding practices stand as an enduring marker of human manipulation of NAPT environment. Each form deserves separate consideration, and each location differed somewhat.

5.3.4 The Manipulation of Hydrology

While not universal, a common characteristic of these hierarchical agricultural societies was the manipulation of freshwater hydrology. Most frequently this took the form of irrigation, but there are several other important forms of water acquisition, storage, and distribution. The simplest form of irrigation, and probably the most ancient, is bearing water to individual plants in some kind of a vessel. However, the redirection of channel flow, and the systematic extraction of groundwater are both much more widely practiced and have a much greater impact on the physical and the social environment.

The most sophisticated methods of irrigation entail complex systems of capture, storage, transport, and release of water. The social structures required to support arrangement were sufficiently intricate to be defined by Karl Wittfogel as "Hydraulic Civilizations". Wittfogel proposed the notion that there was a "hydraulic revolution". He notes that there are several basic requirements for agriculture: temperature, the lay of the land, fertile soils, cultivable plants, and water. Of these only water "In its agriculturally most precious occurrence – as the water of rivers and large streams in arid or semiarid regions – it therefore defied the small-scale approach, which under preindustrial conditions, was so eff in the treatment of soil and plants. In order to bring fertility to large water-deficient areas by the management of substantial sources of water-supply, man had to create large-scale enterprises that usually were operated by the government" (Wittfogel 1956: 153). Central to Wittfogel's argument is the observation that in order to undertake engineering works over large areas required some specialized skills and social organization. A good understanding of basic physics and civil engineering was required to construct effective and enduring irrigation systems. This in turn demanded research and experimentation, and a system of record keeping and knowledge transfer so lessons learned could be transmitted between generations. A highly organized system of specialized labor was required to successfully construct interconnected systems that would work in concert. Hierarchical organization was needed to both direct and successfully complete the work. Good calendars were needed to keep track of work undertaken in a routine fashion at widely separated places. A large number of people had to be forced to take on the heavy lifting of construction and maintenance. Surplus had to be produced to support those who did not farm. Technical details varied from place to place – Wittfogel is quite clear: "Within the orbit of immense cultural variety differences occur" (p. 153).

Many commentators have noted a significant number of examples that one way or another refute some of Wittfogel's more sweeping claims, particularly about the details of social organization, and the economic structures arising from hydraulic civilizations (Graeber and Wengrow 2021). However, despite these differences these kinds of complex feudal and imperial agricultural empires had some remarkable similarities from place to place and existed in some form on both sides of the Atlantic. The earliest evidence so far found is from Mesopotamia, from around 5000 BP. Over the course of the first 3000 years of the Meghalayan other complex, populous, and hierarchical agricultural societies emerged throughout the Mediterranean periphery, and independently emerged and expanded outwards from Central America into both North and South American continents (they also independently emerged in a number of parts of Asia, but those developments are beyond the scope of the current work).

The myriad technical innovations deriving from the development of these empires are visible throughout the regions where they emerged. Most particularly these centered on mathematics, civil engineering, calendars, record-keeping, and surveying. In all cases the key element of underlying these things was the production of significant surplus, to support specialized labor. In the context of the current analysis the most important characteristic of these societies is their extraordinary ability to transform landscapes. Once the logistical details of labor management were mastered the resulting impacts were impressive. Thus, evidence of human labor dating from this time was remains visible from space. Not only is the geographical scope of these changes impressive. It is also worth remarking on the extraordinary multi-generational vision of the planners and builders of the works. The "grand schemes" of terrace-makers, canal-diggers, aqueduct designers, and road builders required an extensive understanding of both time and space. There are, or course, some spectacular mistakes visible in the landscape. Doubtless many lessons were learned at a high cost of trial and error.

In other cases, outside the realm of deliberate manipulations by the people of these civilizations, there were still measurable impacts on hydrology associated with humans. Some were planned, others were incidental. The planned involve small and localized methods for capturing, storing and transporting water. Rainfall capture methods at the farm or village level typically involved fairly small catchment areas, and underground storage. These small and fairly simple systems proved remarkably robust. Some, such as in parts of the Levant were initially constructed in the Neolithic and survived to the present day. Other small-scale methods of irrigation included induced condensation, where ventilated stone mounds served as condensation surfaces, exploiting diurnal variations in temperature to extract and concentrate water from the atmosphere methods. This method was practiced in high altitude regions of the Americas, in particular.

5.3.5 The Effects on Geomorphology

One of the fundamental changes that occurred as a result of these shifts was the emergence of a clear shift in landscape dynamics. The geomorphology of an increasing area of the NAPT has been fundamentally changed by the advent of various forms of agriculture. Was transformed to a similar extent during the Meghalayan. The wholesale anthropogenic reconstruction of three biomes, cleared forests, managed pastures and arable land changed sediment budgets, and the biogeochemical cycles of soluble materials. Some of these changes can be deduced from first principles. In other cases, there is clear empirical evidence visible in the landscape. Undoubtedly there are many places in which the research is yet to be undertaken to understand how landscapes were transformed by the spread of human economic activity across landscapes.

Working from first principles it is fairly straightforward to use contemporary observations of the impacts of, say deforestation, to understand how sediment budgets changed when forests were cleared either to provide wood, or to open farmland. There are substantially different geographical patterns of this process through the NAPT. In the temperate regions of North America through most of the Meghalayan a pattern of what is called agroforestry prevailed. The general categories outlined by Bill Doolittle in his book "Cultivated landscapes of native North America" provides a useful framework within which to consider these ideas (Doolittle 2000). Doolittle proposes horticulture, rainfed, dryland, and wetland systems of cultivation all coexisting, and in many cases intermingling on the North American continent. There are intermingled subcategories that extend from, and include, continued gathering of truly wild species, protection and encouragement of wild species, and the cultivation of domesticates. Within all of these variations it is worth noting that North America was if not the foremost "breadbasket" of the world, most certainly a significant contender for that title. As Doolittle argues, North America was "a major contender for that title" (p. 3). In other words, agricultural productivity on the western side of the NAPT was comparable to that of the eastern side. The implications of this intensity of cultivation show up in a number of different ways. As already noted, when this system collapsed it resulted in a period of reforestation which may well have significant and enduring global climate impacts. In the context of this section of the discussion the role of this type of agriculture in altering sediment budgets is more important. There is not a large literature on this topic. Not a lot of time has been devoted to exploring pre-Columbian soil erosion attributable to land clearance. What is available suggests that the impacts were

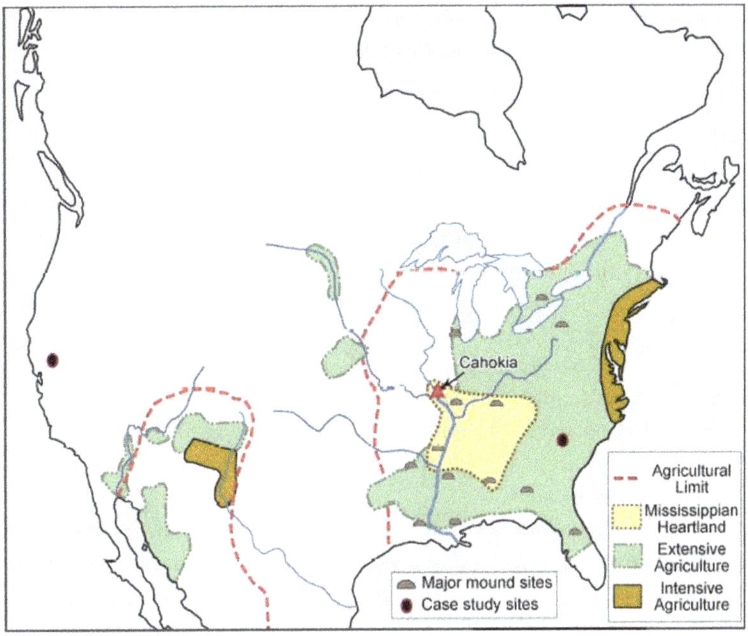

Fig. 5.3 Precontact agriculture and cultural features (After James 2011)

certainly dwarfed by the effects of post-contact agriculture, at least in North America. The extent of agriculture exerted some influence on this tendency. Figure 5.3 shows the probable limits of farming and gives an indication of its intensity (James 2011).

Within the realm of the invention of agriculture, including the Levant, China, and Central America, North America, a significant part of the NAPT, stands as a major region of innovation and development. Consequently, the transformation of the geomorphology of this region started early and endured. Although the effect was relatively slow, playing out over millennia, it was still significant.

5.4 Culture

In parallel with changes in climate there was a qualitative shift in human social organization over the course of the Meghalayan. By the end of the Northgrippian the basic structures of complex hierarchical societies had already been well-established. There is no generally agreed-upon clear human historical marker matching the 4.2 k event. This noted, there is an increasing body of evidence suggesting that matches the replacement of the Egyptian Old Kingdom by the Middle Kingdom (Ran and Chen 2019). Several other broad trends fall within the same general timeframe. In the mid latitudes of the eastern side of the Atlantic the last outposts of Neolithic life were incrementally replaced, as societies supported by Bronze and then Iron smelting technologies spread outwards from the "cultural hearth" of the Eastern Mediterranean (Blanco-González et al. 2018). This change extended into subarctic Northern Europe. In the Arctic there are clearer associations in time – around 4500 BP Greenland's first know human inhabitants arrived, the Independence I in the north and the Saqqaq people in the south. This move has been credibly linked to the climate changes occurring around the boundary between the Northgrippian and the Meghalayan (D'Andrea et al. 2011). In South America, as in Europe the time seems to have been one of incremental spread of urban and hierarchical agricultural civilizations from the cultural hearth of central America. The two prevailing cultural characteristics of the Meghalayan are the final expansion of human habitation into all habitable environments, and the rise to pre-eminence of organized and state-based agriculture as a mode of production. Accompanying this general trend was a consistent overall rise in human population, and a matching growth in the reach of human impact on landscapes.

The most important change was a wholesale shift in the prevailing mode of production. This process is well summarized by Neil Roberts in his book "The Holocene: An Environmental History". He notes:

> From a common starting point in peasant farming there evolved, during the second half of the Holocene a diverse range of cultures based on agriculture and/or pastoralism. The most important of these were socially stratified and involved Asiatic, Classical, or feudal modes of production. All of these shared in common the production of an economic surplus by agricultural labor that was taken and used by another class. This new form of social relationship also changed human-environment relations in several important ways. It meant, for example, that maximum economic output tended to replace the mix of social and economic goals that characterize communal peasant farming societies. Agricultural output was maximized by improving the productivity of existing farmland or by increasing the cultivated area. The latter led to clearance of woodland and draining of marshland, but also included cultivation of marginal land susceptible to soil erosion and other forms of damage. The advent of complex agricultural societies distanced and often weakened the link between people and nature. Nature became less the 'habitat' for the farmer than a set of economic resources to be managed and manipulated by the controlling group. (Roberts 2013: 235)

Roberts' argument is that while farming remained the predominant survival strategy there was widespread change in the social structures within which it was conducted. As a result of these changes there was a fundamental shift in the ways in which the practice of agriculture influenced landscape processes. The resulting transformations can be best understood by exploring the ways in which they influenced ecosystems, hydrological systems, and earth surface processes. It is noteworthy that at this time the distinction between humans and nature, already somewhat porous, become completely permeable. In the Meghalayan it is impossible to model any terrestrial ecosystem anywhere in the NAPT without considering the presence of humans. In contrast, many marine ecosystems, while impacted by human actions, still remained largely untrammeled until the last few centuries of the period.

5.5 The End of the Meghalayan

The end of this chapter marks a turning point in my discussion. One of the central tenets of this book is that the Holocene is ending. Necessarily this coincides with the end of the Meghalayan. As in previous chapters, and more importantly, as in the end of previous ages, periods, epoch and so on, I have stuck to the chronostratigraphic principle that the bounding stratotype is what marks the end of a time slice. Within a geological timeframe this means that the boundary event matches the great transition that moves humanity from agricultural to industrial society. Two or three hundred years is not very much time in the grand scheme of things. Therefore, I have moved that discussion into the next chapter.

As I have noted previously, it is here that I part company with the proponents of the Anthropocene, who are arguing that we are already in the new age. I am not alone in making this argument, others have recently argued for the Anthropocene to be defined as an event, at least comparable to the 8.2 k 4.2 k events (Gibbard et al. 2022). In keeping with current practices, I contend that it is only possible to define and describe the new age when it concludes. There has to be a big enough change in the whole planet's process regime to make it possible to identify the unique characteristics of the new time slice in the geological record. This chapter ends here because the boundary event is sufficiently large and complex to require an entire chapter of its own. As with almost all other boundary events (meteorite strikes aside) it does not have a single neat point in time. Recall, the chronostratigraphic community require the identification of a "golden spike". However, this again can only be identified in retrospect. For now, we must content ourselves with finding the times when key process dynamics underwent dramatic directional shifts, in graphic terms, the bifurcation points of tracking statistics. Some of these are quite easy to identify. The celebrated "hockey-stick" graph of temperatures, and the matching graphs of atmospheric carbon and sulphur is a case in point. Others are harder to pin down. For example, dramatic shifts in the population dynamics of oceanic species are less well studied. This complex pattern of change is the subject of the next chapter in which I will do my best to map out the boundary event that marks the terminus of the Holocene.

References

Bampton M, Kelley AR, Kelley JT (2018) The hyperlocal geography of climate change impacts: long-term perspectives on storm survivability from the Shetland Islands. Hist Geogr 46(1):129–150

Blanco-González A, Lillios KT, López-Sáez JA, Drake BL (2018) Cultural, demographic and environmental dynamics of the copper and early bronze age in Iberia (3300–1500 BC): towards an interregional multiproxy comparison at the time of the 4.2 ky BP event. J World Prehist 31(1):1–79

Bradley R, Bakke J (2019) Is there evidence for a 4.2 ka BP event in the northern North Atlantic region? Clim Past 15(5):1665–1676

Broecker WS (2001) Was the medieval warm period global? Science 291(5508):1497–1499
Butzer KW (1988) Cattle and sheep from old to New Spain: historical antecedents. Ann Assoc Am Geogr 78(1):29–56
D'Andrea WJ, Huang Y, Fritz SC, Anderson NJ (2011) Abrupt Holocene climate change as an important factor for human migration in West Greenland. Proc Natl Acad Sci 108(24):9765–9769
Doolittle WE (2000) Cultivated landscapes of native North America. Oxford University Press, Oxford
Gasse F (2000) Hydrological changes in the African tropics since the Last Glacial Maximum. Quat Sci Rev 19(1–5):189–211
Gibbard P, Walker M, Bauer A, Edgeworth M, Edwards L, Ellis E, Finney S, Gill JL, Maslin M, Merrits D, Ruddiman W (2022) The Anthropocene as an event not an epoch. J Quat Sci:1–5
Graeber D, Wengrow D (2021) The dawn of everything: a new history of humanity. Penguin, London
Helama S, Jones PD, Briffa KR (2017) Dark ages cold period: a literature review and directions for future research. The Holocene 27(10):1600–1606
Jackson RC, Dugmore AJ, Riede F (2018) Rediscovering lessons of adaptation from the past. Glob Environ Change 52:58–65
James LA (2011) Contrasting geomorphic impacts of pre-and post-Columbian land-use changes in Anglo America. Phys Geogr 32(5):399–422
Koch A, Brierley C, Maslin MM, Lewis SL (2019) Earth system impacts of the European arrival and Great Dying in the Americas after 1492. Quat Sci Rev 207:13–36
Ljungqvist FC (2010) A new reconstruction of temperature variability in the extra-tropical northern hemisphere during the last two millennia. Geogr Ann Ser B 92(3):339–351
Mann ME, Zhang Z, Hughes MK, Bradley RS, Miller SK, Rutherford S, Ni F (2008) Proxy-based reconstructions of hemispheric and global surface temperature variations over the past two millennia. Proc Natl Acad Sci 105(36):13252–13257
McKay NP, Kaufman DS, Routson CC, Erb MP, Zander PD (2018) The onset and rate of Holocene Neoglacial cooling in the Arctic. Geophys Res Lett 45(22):1–10
Ran M, Chen L (2019) The 4.2 ka BP climatic event and its cultural responses. Quat Int 521:158–167
Roberts N (2013) The Holocene: an environmental history. Wiley, London
Sáez A, Godfrey LV, Herrera C, Chong G, Pueyo JJ (2016) Timing of wet episodes in Atacama Desert over the last 15 ka. The Groundwater Discharge Deposits (GWD) from Domeyko Range at 25° S. Quat Sci Rev 145:82–93
Spielhagen RF, Werner K, Sørensen SA, Zamelczyk K, Kandiano E, Budeus G, Husum K, Marchitto TM, Hald M (2011) Enhanced modern heat transfer to the Arctic by warm Atlantic water. Science 331(6016):450–453
Wittfogel KA (1956) The hydraulic civilizations, Man's role. In: Thomas WL (ed) Changing the face of the earth. University of Chicago Press, Chicago

Chapter 6
The Anthropocene Boundary Event

The Present and Beyond

Abstract The end-Holocene boundary event that equals or exceeds the 8.2 k Event and the 4.2 k Event. The environmental markers of a change of this magnitude including enduring changes of atmospheric and oceanic chemistry, and climate. Widespread reorganization of ecosystems, accelerating rate of extinction. A qualitative transformation of all major Earth-surface systems.

Keywords Boundary event · Transformation · Extinction · Atmospheric chemistry · Climate change

6.1 Introduction

This chapter is a discussion of the End Holocene Event. In this book I have argued that designating the present time as the Anthropocene, the "Age of Humans", is premature. All previous geological divisions at all scales have been designated retroactively. It is useful to remember that they are a scientific convenience: a useful intellectual tool to summarize the complexities and scope of the 4.7 billion years of geological time. Each division and sub-division uses a set of environmental markers to lump together predominantly consistent process regimes in both the ecological and the geological worlds. The catastrophic ending of one time unit sets the stage for the beginning of the next. But it is only in retrospect that this new unit can be named and characterized.

So what evidence is there that the Holocene are now at an end? The evidence that will show up in the ice core record, should it survive, or otherwise, in the sedimentary and speleothem records, is the final temperature "spike" of the ζ-curve. This sharp directional change in global temperatures is now of the same magnitude as a the 8.2 k event. It is rapidly approaching the magnitude of the end of the Last Glacial Maximum. However, this is not the only indicator of systemic process change, rather it is a useful proxy which gives us a window on changes in almost every other area of Earth system process. In fact, the only Earth surface systems which are not undergoing a significant shift in process dynamics are the long-term

M. Bampton, *The North Atlantic Polar Triangle*, Springer Polar Sciences,
https://doi.org/10.1007/978-3-031-27264-6_6

tectonic trajectories of continental plates, and the internal crustal dynamics driving them. High magnitude changes are occurring in all other key surface systems. Purely geophysical systems such as the atmosphere are undergoing changes in both composition and dynamics. Organisms in every known ecosystem are transforming, with extinctions, evolution, and behavior all in dramatic flux. The physical composition of many Earth systems are likewise changing rapidly, as novel materials are introduced into an ever-increasing variety of environments. In some cases, this is simply a question of existing materials, such as free carbon, finding their way into environments like the Arctic Ice, where they were little known previously. In other cases, the materials themselves are wholly novel. Thus, long-chain polymers did not, as far as we know, exist in nature, prior to their invention. However, plastics, micro-plastics, and nano-plastics suffuse the entire surface environment of the Earth, and most of the organisms that live there.

Some of the other systems showing "spikes" in change rates are causally linked to the ح-curve, and when traced over the course of the Holocene have a similar rise, plateau, fall, and then jump. Others work on different temporal scales, and have long periods of incremental change followed by other shifts. Thus, human population has gradually incrementally increased over its 200,000-year history and is now spiking. Species diversity, and total biomass has fluctuated on several different timescales, crashing during major time changes, and then reviving dramatically. Now, however, extinction rates are rapidly accelerating, diversity is crashing, and biomass is shifting markedly away from larger and more complex organisms, towards small simpler structures. Total ice volume crashed at the beginning of the Holocene, then stabilized until the 8.2 k event, and then slipped to a lower equilibrium. There were several small fluctuations around a mean value for the next eight millennia, and now the volume is rapidly decreasing again. A selection of these curves gives an indication of the general pattern and reveals several dissimilarities in form and scale. All have in common a directional change that starts at some point in the past 500 years. In all cases this directional change marks a change in a raw value, and a change in rate. That is, the changes that are happening at present represent both a quantitative and a qualitative shift in the way most Earth systems are behaving.

To organize my discussion of this moment of change I shall draw on the tradition of physical geography and divide the world into a series of interconnected "spheres": the geosphere, the atmosphere, the hydrosphere, the cryosphere, and the biosphere. Each of these has a distinctive directional change analogous to the average global temperature spike. In some cases, the spikes are increases, in other cases they are declines. However, a common characteristic prevails in each sphere: numerous subsystems have changed, and the rate of change has changed within the past three or four centuries. In some cases, these shifts can be directly measured. For example, in the biosphere the area of land monocropped has increased exponentially. Meanwhile in the cryosphere the area of ice has decreased in a similarly dramatic fashion. In some cases, the rate of change is more telling than a raw number. The rate of extinction has changed upwards. The rate of desertification has increased upwards. The rate of deforestation has similarly increased.

Documenting this widespread set of shifts across the five spheres of earth surface processes will allow me to argue that we are in the midst of a chronostratigraphic boundary event. Available evidence indicates that the change is profound, systemic, ubiquitous, and directional. The industrial rhetoric of the nineteenth and twentieth centuries was a model of clarity – all of nature was to be tamed and bent to the will of humanity. There is abundant evidence to indicate that while these two concepts are conditional at best, humanity's collective transformational impulse has now come to fruition. Every Earth surface system is now in the process of changing.

It is highly likely that many of the trends currently underway are irreversible: we probably no longer have the opportunity to stop many of the systemic transformations. I cannot imagine a global political or economic transformation of sufficient magnitude to halt, let alone reverse, the process dynamics currently underway. Our work now must be to strive towards some method of steering the changes and managing the consequences. Avoiding catastrophic human population collapses, for example, would be a desirable outcome. If we learn only one lesson from Earth history it must be that dramatic change is not pretty or pleasant. We have passed many points of no return. Extinct species cannot be revived, and even if they were, they could not be woven into the fabric of existing ecosystems. There is no realistic prospect of extra-planetary colonization. We have leapt into the darkness. What distinguishes this time from all previous periods of the human experience is that we know what we have done, and we can predict with some certainty where we are going. Details remain to be worked out, of course. But the overall trajectory is established. What remains is for us to chart a course through into the next hronostratigraphic unit. Whether this is a transition into a new age, a new period, or a new eon remains to be determined. We are not entering the Anthropocene age rather we are the Anthropocene boundary event.

6.2 Atmosphere

Atmospheric variables serve as a useful indicator of other process dynamics. That is, the atmosphere responds rapidly, and in a some readily measurable ways to other shifts in environmental processes. This is sometimes a source of confusion, as people unfamiliar with these kinds of measurements sometimes confuse weather – what is happening right now – with climate – the broader historical and geographical trends of atmospheric behavior. Further problems ensue because science denying rascals and charlatans are quite happy to exploit this confusion for their own nefarious ends. The argument usually boils down to some version of the assertion that: "there is no such things as global warming because it is cold today". The rapid reaction of the atmosphere to changes in Earth's energy budget is because it is the conduit through which nearly all of Earth's energy passes, often multiple times. As such it is an extraordinarily sensitive monitor of the minute-by-minute fluctuations and shifts in process. The extraordinary interconnectedness and reactivity of atmospheric processes globally, and with numerous other systems, is so complex that it defies simple linear mathematical modelling. While the physical laws governing the behavior of the atmosphere are well established, and mostly date back to the eighteenth century, it turns out that the complexity of their operation at a global scale presents significant modelling challenges. Sometimes quantity is a quality.

It is typically described using four metrics: temperature, pressure, humidity, and composition. Each of these are interrelated, and each gives us a window on many other changes. The combined gas laws tell us that pressure, temperature and pressure are all closely related. PV/T = K (pressure multiplied by volume divided by temperature equals a constant). Thus, if any one variable changes, the others change proportionally to accommodate the shift. If atmospheric temperature changes, pressure and volume adjust in a predictable manner. Likewise, if either pressure or volume changes the remaining two variables also change. The laws of thermodynamics allow us to understand how atmospheric humidity interacts with these three other variables. The composition of the atmosphere varies from place to place, both vertically and horizontally. The majority of its components are pretty stable. 78% Nitrogen, 21% Oxygen, 0.93% Argon, 0.04% Carbon dioxide, and trace amounts of neon, helium, methane, krypton and hydrogen. This leaves just 0.03% unaccounted for. However, there are significant variations in this 0.03%. Differences in the amount of water vapor and the other "trace" components all translate into differences in composition. Each of the four measurable atmospheric variables can change very rapidly indeed. In almost all circumstances they respond to external influences more rapidly than any other environmental variable. Over the past 300 years all four have undergone significant changes.

Easiest to track by far is the change in temperature. Atmospheric temperatures is one of the most widely sampled of environmental variables, and it is also has one of the longest set of direct measurements. The first instrumental measurements were made between 1654 and 1670 within the "Medici Network", in northern Italy (Camuffo and Bertolin 2012). This effort, sponsored by the Medici family, yielded a set of records documenting a brief and spatially constrained data series. However, it was followed by a fairly rapid increase in measurements. The Meteorological Observatory Hohenpeißenberg, in Bavaria has a continuous record dating from 1780 to the present (Deutscher Wetterdienst n.d.). By 1767 periodic observations were made at the Green Templeton College of Oxford, and a daily schedule of recording was established by 1815. Within a comparable timeframe most other European nations started to gather weather data. By the mid-nineteenth century the European colonies in Africa and the various colonies and nations of the Americas had established regular temperature measurements. The curve for Hohenpeißenberg essentially represents the trend. Measurements started in the latter part of the Little Ice Age, and by the time a regional network was established there was a slow but steady trend of warming. This continued until around the beginning of the twentieth century, at which point the curve began to trend upwards. The trend accelerated until around 1950, which is frequently cited as the inflection point. At this point the gradient of the curve trended sharply upwards and yielded the definitive "hockey stick" graph.

One of the defining characteristics of the end of the Holocene is this rapid shift in the temperature of the atmosphere. Given the close relationship between temperature, pressure, and humidity, some reasonable deductions can be made concerning other atmospheric variables during the same period. This is useful, as there is a much shorter record of direct measurement of these other meteorological variables. While pressure records are highly variable, and difficult to track a good indicator of the changing pressure characteristics of the atmosphere is in storm tracking. It is evident that there has been a marked increase in short-term storm frequency during this period. Tracking the changes in long-term storm frequency is more challenging. However, there are some useful proxies available, that give some indication of the patterns Foremost among these is the GISP2 Greenland Ice Core project, which gives some insight into North Atlantic storminess for the entire Holocene. This is not a definitive picture, but it presents some intriguing patterns. There is a clear indication that following the end of the Little Ice Age there was a marked trend upwards in overall storminess. Using a number of other proxies, including oxygen isotope records, tree-ring data, and historical records there is clear indication that the overall pattern of storminess in the NAPT underwent a significant transformation during this time. The best indicators of this tend are the Atlantic Meridional Oscillations, including the North Atlantic Oscillation. These long-term shifts in pressure patterns and their linked effects on ocean temperatures have been modelled back to 18,000 BP (Repschläger et al. 2015). The current interconnections of the system are illustrated in Fig. 6.1.

Some aspects of atmospheric composition are challenging to monitor. Humidity is, like pressure, sufficiently ephemeral that it presents some challenges in the pre-instrumental period. However, a combination of proxy measures, models, and more recently direct measurements give us some strong indications of how current patterns are changing. Other characteristics of composition are much better monitored than other variables, particularly for much of the Meghalayan. This is primarily as a result of the work from the Ice Core projects. In the current context the Greenland cores are by far the most useful, though other glacial cores form within the region also reveal useful information. As previously noted the Greenland ice cores provide a continuous record of 123,000 years – a large proportion of the entire time anatomically modern humans have been present on the surface of the Earth. There are several chemical "signals" held within this record. I have already noted that the salt record can be used as a proxy for storminess.

The two most important gasses in terms of climate change, methane (CH_4) and carbon dioxide (CO_2), are reliably retrievable from the air bubbles trapped in the ice cores. As with other proxies they show a clear pattern of fluctuation around a mean value for this entire period. In fact, the last major fluctuation in either CO_2 or CH_4 was at the beginning of the Holocene. CO2 follows the ζ-curve. Methane presents a mirror image of the ζ-curve's general form. There is a sharp rise at the beginning of the Greenlandian. This is followed by a gradual decline through until the middle Northgrippian, and then a gradual upward trend, ever steepening, until the present. Once again, an upward inflection characterizes the end of the Meghalayan, but it is not the "spike" that we have seen in so many other cases. Rather it is an increasing

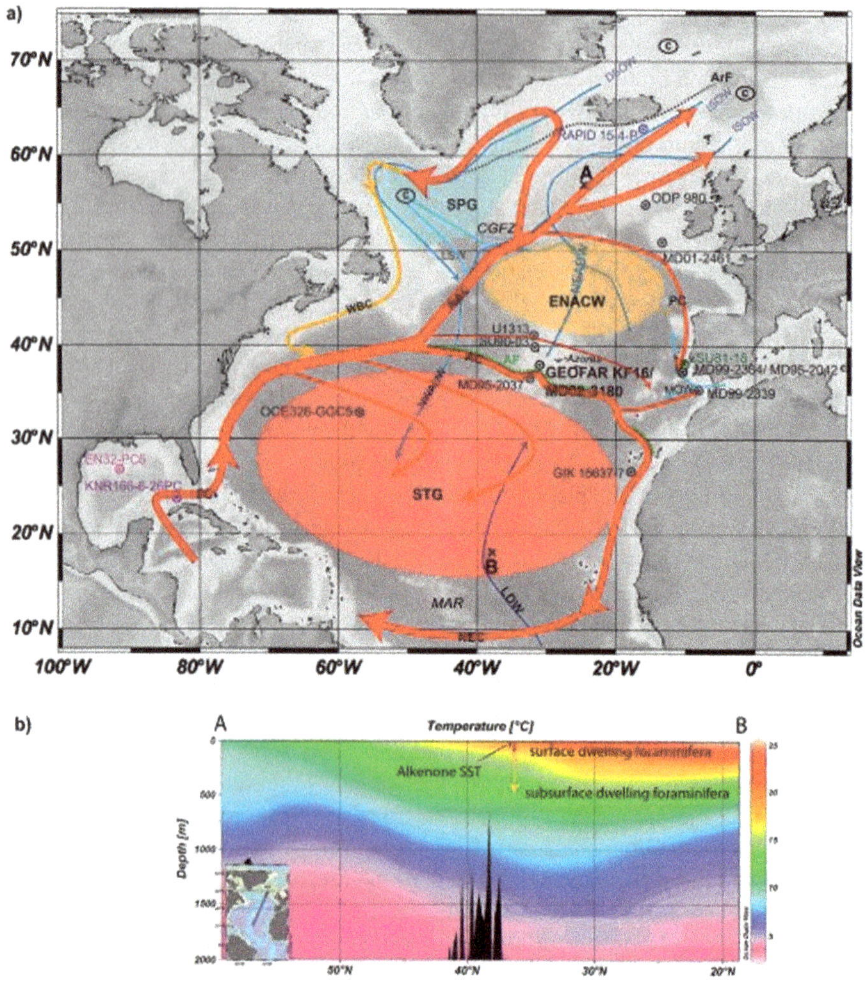

Fig. 6.1 NAO and AMO interactions with prevailing pressure structures in the North Atlantic (After Repschläger et al. 2015)

upward curvature. There are several drivers of this change. Methane is the product of two major set of processes, ruminants and melting ice. The decline from the early Greenlandian, to a low in the middle Northgrippian is the product of declining methane releases, as the atmosphere reached an equilibrium, following the massive releases caused by the melting of Ice Age permafrost. The gradual rise from the middle Northgrippian onwards can be attributed to the spread of cattle-farming, as agriculture became an increasingly important part of human economies, and also to the gradual establishment of large grazing populations of other herbivores. Some of these were probably partially managed by humans, such as the Plains buffalo of the Americas, and the Arctic reindeer of northern Scandinavia. Others were predominantly wild, such as the migratory herds of caribou of Arctic America. A second contributor to the rising quantity of methane in the atmosphere is the recent acceleration in the melting of permafrost associated with the current warming trend. The overall shift in global methane can be reduced to a change in the relationship between carbon-fixing organisms and their metabolization. That is, the organisms that ingest and process carbon and how they do so. The shift in the classical fox/rabbit/grass illustration is entirely transformed.

6.3 Cryosphere

Cryosphere changes fall into four categories:

1. Volume & Mass
2. Location
3. Quality
4. Duration

Volume and mass are the net amount of ice in the world. There are a number of reliable calculations of the total amount of ice in the world. There are also good estimates of where it is located, and how it fluctuates seasonally and over the long term. Both volume and mass have undergone marked change throughout the Holocene. The current rate of change is rapid, but it not unprecedented. There have been two major instances in the quaternary where global ice has undergone a change of comparable magnitude: the end of the LGM, and the 8.2 k event. There is a yearly fluctuation of ice quantity, but there is also a clearly defined trend. It is in this trend that the spike is visible.

The location of ice has changed in several ways. For most of the Holocene there have been continuous ice masses somewhere at almost every latitude from the equator to the pole. At present only one glacier straddles the equator, the southernmost tongue of Equador's Volcán Cayembe (00° 00′00″ 77°59′24.58″W) (Fig. 6.X The Heart of the Andes). Between 35°N and 35°S, glaciers occur only at high altitudes, such as in the Himalayas, Andes, East Africa, Mexico, New Guinea and in the Zagos mountains of Iran. There are no glaciers or mountain ice caps immediately around either of the Tropics between latitudes 24° and 19° north and south. Nor is there any permanent ice between Latitudes 56°S and 61° S, where the Southern Ocean forms a continuous uninterrupted seaway through all 360° of Longitude. Above the Arctic Circles there are permanent glaciers and ice-sheets on many land surfaces, and on nearly all mountains. Seasonal ice covers much of the rest of the land and the ocean during winter months. Semi-permanent ice is present in many mountain regions at lower latitudes and can be found in such unexpected places as the Cairngorm Mountains of Scotland, and the White Mountains of New England. Antarctica has the oldest ice sheet with some parts over 2.7 million years old. In Greenland the ice is around 130,000 years old. Some of the tropical ice sheets are significantly younger, for instance Mt. Kenya has ice that dates back to 6000 BP (Karlén et al.

1999). Consistently the ice predates the LGM. Ice is now melting rapidly, and this global distribution is irrevocably transforming. Equatorial ice will be gone by the end of the twenty-first century. The Arctic will be ice-free during the summer, and most semi-permanent snow patches will be gone within the same timeframe. Most, if not all sub-arctic glaciers are vanishing at a comparable rate and will probably be gone by the end of this century, or early in the next. The question that remains is how much of the two major continental glaciers, in Greenland and Antarctica will survive.

There are four aspects of ice quality that have changed in the Holocene: temperature, pressure, chemistry, and color. The impact of changing ice temperature is in two areas, the body of large ice masses, which have become increasingly porous and fragile, as they warm towards 0°. The second area of change is at the interface between ice and other media. This the air-ice interface has undergone a significant warming in the past century. Likewise, the ocean-ice interface has significantly changed. These two interface changes have, combined, produced a situation in which the overall thermal mass of ice sheets and glaciers large and small is higher. Consequently, the third interface, between ice and bedrock is more frequently warmer than it was earlier in the Holocene. Historically there was a distinction between "cold" glaciers, frozen to the underlying bedrock and "warm" glaciers, frozen to the bedrock. This distinction is becoming increasingly irrelevant as the Earth's last "cold" glaciers warm up.

Changes in ice chemistry are better documented than many other areas of change as ice cores have preserved measurable evidence of sodium, carbon, sulphur, and nitrogen compounds (among others) are preserved throughout the record. They are directly linked to changes in atmospheric chemistry as ice chemistry is determined either by the composition of precipitation or dry deposition of other materials. Throughout the Holocene there have been periodic fluctuations in this record, mostly driven by vulcanism. Key periods of volcanic activity, such as the (geologically) recent 1815 eruption of Indonesia's Mount Tambora, show up as an increase in ice-core sulphates and chlorides (Udisti et al. 2000). The peaks and troughs of volcanic particles are driven by long-term tectonic processes that fall outside any Earth surface process causation if which we are currently aware. They are an independent variable and, as previously noted, can exert significant influence on climate. Within the Holocene, and indeed throughout the Quaternary these values periodically spiked, and the declined again, as individual episode of volcanic activity great and small erupted. In contrast there are key historical markers attributable to specific human actions, and a directional shift in ice chemistry associated with the onset of industrialism. One key historical marker is the increasingly carbon-rich ice layers associated with the advent of widespread coal-combustion, following the broad adoption of steam power and the explosion of iron and steel smelting in the early nineteenth century. Another is the layer of radionuclides associated with the advent of testing and use of nuclear weapons and subsequently nuclear power in the mid twentieth century.

Such compositional markers show up as constrained layers within the ice, as they have a known onset, and then a documented decline as technologies and practices change. Other chemical contaminants associated with industrialism have consistently increased over the past 500 years or so. Thus sulphates, methane, CO_2, and hydrogen dioxide have all steadily grown. These, when viewed in the perspective of the past 1000 years show the "hockey stick" form disseminated in Michael Mann's work. Viewed over the longer perspective of the entire Holocene they take on more complex forms. In some cases, coming from "out of the blue" as new synthetics show up in the chemistry, and in other cases fluctuating around low-level means, with occasional anomalous spikes and declines, and then steadily trending upwards as industrialism becomes prevalent.

Pressure changes are associated with the overall loss of ice mass. As ice is lost, the overall internal pressure of ice is transformed. This affects temperature, melting rates, and plasticity. This in turn affects the dynamic behavior of glacial ice, and the integrity of ice sheets (British Antarctic Survey n.d.).

The color of ice is determined by three factors, age, the dry deposition of atmospheric pollutants, and proportion of ice to ice-burden. Ice-burden is the material carried in the ice, in the case of glaciers the various materials are lumped together under the term "till". This is a combination of materials picked up as ice passes over bedrock, eroded valley sides, rock fragments that tumble onto glacier surfaces. As melting, particularly at high latitudes, has accelerated the proportion of old blue ice to newer whiter ice has shifted. Dry deposition of combustion byproducts has periodically changed the surface color if ice and snow. This, as with ice chemistry, is cyclic. On occasion natural processes, particularly vulcanism can impact the color. In the late Holocene the advent of coal burning, and more recently widespread forest fires has impacted the color. The change in the proportion of ice to ice-burden similarly affects the color of ice, as the proportion of burden increases the radiation response of ice is changed. The collective consequence of these shifts is to move ice from a white material that reflects and transmits a high proportion of incoming short-wave solar radiation to a darker material that tends to absorb a high proportion of incoming radiation. This increased radiation absorption is directly correlated to increased melting.

The duration of ice cover has changed in both terrestrial and aquatic settings. In terrestrial settings this is directly correlated to ice budget in glaciers. Where accumulation exceeds ablation glaciers advance. In all known glaciers the past 100 years has seen a rapidly accelerating rate of ablation and decreasing accumulation. This is attributable to a reduction in snowfall as winters become shorter. Net snow accumulation globally is significantly lower year on year. Likewise, seasonal snowfields have shorter duration. Aquatic ice in freshwater environments is characterized by later freezing, and earlier ice-outs (Fig. 6.X Ice-out in selected Maine Lakes 1800 – Present). In marine environments this is most clearly visible in the reduction in Arctic ice cover. The Northwest Passage is now open during summers for the first time in history, and there is a high likelihood of an ice-free Arctic Ocean for part of the year before century's end.

This is not merely a question of disappearing ice. The changes that will accompany this shift are profound, from global energy budgets, to ecology, to hydrology.

6.4 Hydrosphere

The spikes in hydrosphere processes can be seen in each of the “steps” comprising the classical version of the hydrological cycle as illustrated in every middle school science classroom in the developed world. That is, they include the oceans, terrestrial surface water, groundwater., atmospheric water, and water held in living tissues. Not only has the water resident in these storages changed, the pathways have changed, and the temporal and spatial distribution of water has changed.

Ocean water is experiencing two dramatic changes, temperature and composition. Temperatures are rising rapidly, and at unequal rates. Some locations, such as the Gulf of Maine, for example, are experiencing very rapid warming. This is apparent in the out migration of historically local species, and the in-migration of competitors. Two striking examples of this are the northward migration of the iconic New England lobster, and the matching northward migration of one of the ocean apex predators, the Great White Shark. As previously noted the composition of ocean water has been changing since the advent of agricultural deforestation, as sediment runoff has increased. There has been an exponential upward trend in sediment runoff in the Late Holocene. The most dramatic recent change, however, is the increasing occurrence of synthetic materials. Originally limited to heavy metals, and then radionuclides, the synthetic pollutant load of the ocean has undergone a massive increase as plastics have started to dominate people’s daily lives. Originally confined to the macro plastics of common garbage, this pollutant loading has now started to include micro plastics, and increasingly nano plastics.

Groundwater is undergoing three significant changes. First, as precipitation regimes change with warming climates long-term recharge rates are changing. In those areas of the globe experiencing an increase in precipitation there is a raising of the water table, and a steepening and heightening of flood hydrograph curves. This mean the time lapse between a storm event in a catchment basin and the maximum flow in a stream channel is shorter, and when it comes, the flood peak is higher. In regions where the precipitation is reduced there is a general lowering of the water table. To put it in plain terms, the well runs dry.

Secondly, the quality of groundwater is changing. In many cases lower water tables translate into changes in the flow rates between groundwater and ocean water. In coastal and island regions as the water table falls there has been a consistent pattern of salt-water incursion, rendering ground water brackish. There is also a change in groundwater quality attributable to a host of other pollutants. Fracking, for example, has resulted in the contamination of groundwater in surrounding watersheds, as the injected lubricants and particle have invaded pore spaces in rocks. Other contaminants include outflow from mining spoil heaps, heavy metals, and human waste materials.

Thirdly, groundwater has declined in those areas where deep pumping has allowed people to “mine” ancient water, left over from the Last Glacial Period, as in the Ogallala Aquifer (USGS n.d.), and the Nubian Sandstone Aquifer System. These are both examples of the global trend towards overall decline and is an instance of an inverted hockey stick curve, in which the end of the Holocene is marked by a downward trend (UNESCO n.d.). There has also been a marked decline in areas where water is recharged in real time, but not as fast as it is extracted. The critical shift in this arena has been the replacement of manual, animal or wind-powered extraction to gasoline-powered extraction. The advent of internal combustion pumps has resulted in the opening of previously arid regions to irrigation agriculture, effectively exchanging petroleum for water.

The surface water step of the hydrological cycle has the greatest longevity of deliberate human modification. As noted in Chap. 5, Karl Wittfogel uses the term "Hydraulic Civilization" to characterize ancient agricultural empires such as Egypt in which control of flowing water was a central component. In this model social control was necessary to marshal labor resources to build the structures required for the management of water. These included dams, reservoirs, aqueducts, canals, and cisterns in complex and carefully engineered linked systems. The scope of such works is impressive, and many have survived in working order for thousands of years. However, the Industrial era ushered in a new level of hydraulic control. Initially the primary drivers of this effort were the need to control water for transportation, and for power. However, increasingly complex systems have evolved over the past three hundred years. At present less than 1% of all rivers in the US are designated as "wild" – that is, free from impoundment, canalization, or some other form of regulation (US Department of the Interior n.d.). In other parts of the NAPT, for example Canada, the proportion of wild rivers is somewhat higher (Grill et al. 2019). However, throughout the NAPT it is apparent that the majority of rivers are impounded multiple times between source and sea. This, then is a predominant characteristic of Late Holocene drainage. This change has several impacts. It transforms the ways in which runoff is delivered downstream, and to the ocean. It transforms the downstream delivery of sediment and solutes to oceans. It transforms the exchange of organisms, both as downstream nutrient supplies, and the upstream migration of anadromous fish. This wholesale reorganization of surface hydrology has some immediately obvious impacts, as flood patterns change, and as sediment and fishery movements are transformed. However, longer term impacts are hard to assess. For example, sediment delivery times within large river systems can take centuries, or even millennia to play out. Thus, a dam built now, may result in sediment depletion in coastal regions 400 years later.

Exacerbating the impact of impoundment is the transformation of surface quality. Late Holocene surface are significantly different from early surfaces in two ways. First, there is widespread removal of complex vegetation cover. Deforestation from the equator to the sub-arctic has resulted in a change in the impact of precipitation on soil surfaces, and a decrease in channel delivery times. Rain fall moves across cleared surfaces more rapidly than across vegetated surfaces. Forest clear cuts are particularly susceptible to this transformation. The effect is increased still further in the development of the waterproofed surfaces of built environments. Cities are primarily impervious surfaces, and are designed to shed water as rapidly as possible, immediately redirecting into subterranean drainage systems. This creates an environment sometimes described as "pseudo-karst". That is, they resemble limestone landscapes, where flowing water is quickly redirected into caverns and underground rivers.

Atmospheric water has its distribution in both space and time changed. Global shifts in precipitation are primarily driven by the overall rise in temperature currently underway. More thermal energy results in higher evaporation. The redistribution of atmospheric water influences the hydrological cycle directly. It also has an indirect effect, by changing the distribution of thermal energy. Latent heat, stored in the molecular energy of water vapor, is released when condensation occurs. Dramatic shifts in the quantity of water vapor in any given parcel of air equates directly to a proportional shift in the amount of thermal energy either taken up or released. Widespread drought conditions, as they become increasingly common in sub-tropical latitudes, are exacerbated by the continuing absence of surface water which might otherwise evaporate, so moving some of the excess heat away from land surfaces. In other words, the hotter and drier it is, the hotter and dryer it will become. At a certain point drought creates a positive feedback system.

The two remaining steps of the hydrological cycle, ice and oceans each have their own subsection. However, to complete the model I'll note that there is much less ice. And the oceans are much warmer.

6.5 Biosphere

Transformations of the biosphere are of sufficient magnitude to give us every indication that we are living in a chronostratigraphic boundary event. There are system-level transformations underway at all trophic levels, and within nearly all biomes, terrestrial and marine. Much attention is paid to the macro-faunal changes, and the loss of plant and animal species through extinction. A simple metric is to note that at present about 40% of known plant species are now either extinct or facing extinction (Royal Botanic Gardens, Kew n.d.). On the other hand, it is also important to recognize that some species from microorganisms through insects, to vertebrates, are doing rather well, indeed prospering.

However, biosphere processes are a proxy for systemic changes. Chronostratigraphic boundaries are marked by shifts in conditions so dramatic that some species do not survive, and others must evolve. So, in all chronostratigraphic boundary events there are evolutionary winners and evolutionary losers. There are some general characteristics which distinguish these two. Paleontology teaches us that evolutionary survival is determined to some extent by luck. Surviving a chronostratigraphic boundary event requires extra luck. Being in the right place, at the right time, and having the right set of adaptive evolutionary potentials keeps a species, and thus a genus, a family, and a taxon in the evolutionary game. There are some general principles that work in tandem with blind luck. One of these is that size matters. Evolutionary biologists suggest three rules: "Bergmann's Rule is the tendency for warm-blooded species at high latitudes to be larger than their close relatives nearer the equator. The Island Rule is the trend for small species to become larger, and large species smaller, on islands. Cope's Rule is the tendency for lineages to increase in size over evolutionary time" (Purvis and Orme 2005). During boundary events the tendency is for large organisms to become extinct. Highly specialized species that have very specific environmental and dietary requirements are often vulnerable to change. Species with a limited geographical range are similarly vulnerable.

Conversely the chances of survival are greater for smaller organisms with generalized survival strategies and large geographical ranges have a better chance of survival. At present there are several specific criteria that improve chances of survival that are linked to the conditions currently prevailing in the environment. Species that have co-evolved with humans are frequent winners. Rats, pigeons, cockroaches, human eyebrow mites, and bed-bugs are examples of singularly successful wild species. By sheer count three human domesticates, chickens, pigs, and cows represent a similarly successful enterprise. Other evolutionary winners at present are those microorganisms (including viruses) that have proven successful in exploiting people's behaviors. As I write this the COVID-19 pandemic rages outside. HIV/AIDS, the common cold, dysentery, cholera, and many other pathogens have successfully expanded their range as human population has increased, and as interaction over long distances has become more common.

Within this category of current evolutionary success stories human populations hold pride of place. The total number of humans on the planet is forms undoubtedly a hockey stick curve. The number of people in the world continues to grow exponentially, so far defying Thomas Malthus' prediction. This change undoubtedly serves as a reliable proxy measure for many of the other changes I have described, as it is at the root of nearly all of them. There are two causal structures to this phenomenon. First, in simple arithmetical terms more individuals of any species have a simple linear impact on the environment. More of any organism eat more food, drink more water, and breathe more air. There are several other purely biological characteristics of contemporary humans that are novel. Declining infant mortality has continued to outstrip declining fertility to drive an increase in net human population. An increase in life expectancy has further amplified this trend. Collectively a larger number and a higher proportion of contemporary humans are taller, weigh more, and consequently consume more calories than at any time in the past. The multiplier effect that is most significant in this case is that this greatly expanded population is supported by several key shifts in behavior.

6.6 Culture

Distinguishing the social and cultural characteristics of the end of the Holocene from the human biological characteristics is an interesting challenge. Using the curve models that have provided the intellectual backdrop for the previous discussions most cultural traits fall into either the hockey stick or the inverted hockey stick classes. The purely biological aspect of human engagement comprised the final section of the last section, and it started and ended with the observation that population numbers and biological characteristics comprise hockey stick curves.

Moving beyond this, however, it is possible to identify a series of unique cultural characteristics and trajectories that entwined with the current chronostratigraphic boundary. These have different shaped curves, but as with previous time-dependent trends all currently have a sharp inflection point. Some of these are objects, and some are behaviors. All are unprecedented, either in quantity or quality. They include:

1. Places and connectivity
2. Materials and machines
3. Knowledge and memory
4. Energy

I will consider each in turn.

6.6.1 Places and Connectivity

Human connectivity, the personal physical contact between different people currently occurs at an unprecedented rate. Within the NAPT more people will touch more other people from more different places than at any time in human history. In part this is because there are more people to start with. It is also, however, a result off the concentration of people in urban environments – more people live in cities than at any time in the past. Partly it is because people move over larger distances and with greater frequency than at any time in the past. There are many proxies for this tendency. One of the most accessible is the number o0f air miles travelled. The first manned heavier-than-air flight was probably that of the Wright brothers' 37 m hop on a beach in North Carolina, in December 1903. With a few interruptions for wars, terror attacks, eruptions, and pandemics December 2019, 116 years later, in the USA alone there were 90,371,414 revenue miles of flight (US Bureau of Transportation n.d.). There are numerous other similar metrics which can demonstrate a rapid increase in connectivity between people in the most recent centuries, decades, and years of the Holocene. The characteristics of this increased connectivity are summarized in Milgram's formulation "The Small World Problem" (Milgram 1967). This is one of several landmark discussions of how contemporary social networks are tightly woven. This is manifested in realm of parlor games such as Jewish Geography or Six Degree of Kevin Bacon, where ostensibly extended communities are revealed to be closely connected. Current epidemiology provides a more formal perspective on how increasing interconnectivity functions.

6.6.2 Materials and Machinery

The transformation of materials and machines is twofold. First, there is the matter of novelty. Secondly there is the matter of force multiplication. In the case of materials novelty is worth considering in detail. There are, simply stated, materials that have never previously existed in nature now abundant across the face of the Earth. The most prominent examples of this at the moment are plastics. Synthetic organic compounds include a variety of novel materials from PFCs to long chain polymers. These are completely novel material that have only existed for a little over a century. During this time, they have gone from a chemical curiosity to a universal stratigraphic marker. Sediment samples collected on every continent now consistently show traces of microplastics in measurable quantities. Water samples from every ocean show suspended plastic fragments. Tissue samples from both terrestrial and marine organisms at every trophic level show the presence of nanoplastics. There are plastics present in newly deposited ice on all of Earth's major glaciers. There are plastics in Earth orbit. There are plastics on the moon, and on Mars. And there are plastics dispatched on interstellar journeys.

The list of other novel materials is long, and growing. However, it can be reduced somewhat by categorizing. In addition to polymers there are metals, radionuclides, composites, fractioned hydrocarbons, and pure elements. All of these have come into existence by synthesis, purification, manufacture, or smelting. Some have relatively short lives, however many others have half-lives measured in centuries or millennia.

In addition to the materials that are completely novel, there are those that are now present in much larger quantities than at any time previously. This crude oil and related hydrocarbons have been present on the Earth's surface in small quantities, and in a few locations such as tar pits and oil-seeps for several million years. However, a single event such as the Horizon Deepwater Oil Spill has released several million years-worth of unrefined crude oil into the environment. Perhaps the most important of these multiplied materials is carbon. First in its relatively pure form, as soot. More importantly in the form of carbon dioxide. This, and several other combustion byproducts, including methane, and sulfur dioxide all previously existed in the atmosphere. However, they are now present in sufficiently unprecedented quantities that their signatures are the original progenitors of the "hockey-stick" graph to which we have some frequently referred in this book.

Successful human technical innovations are "force multipliers". Thus, the first ceramics and metals served to increase the impact of human labor on the environment. Successive innovations have all increased humans' transformative abilities. Agriculture, and the subsequent emergence the complex and socially stratified of "Hydrological Civilizations",. Other innovations have been noted in the preceding chapters in which the characteristics of pre-industrial technologies are discussed. However, the defining characteristic of the Industrial Revolution is the exponential increase in human mechanical power. It is not practical to try and review the vast array of devices, systems, and methods that comprise the mechanical innovations of the past 250 years or so, if we mark the beginning of the impacts of the Industrial Revolution from the practical application of the first steam engine. Any such list would certainly be incomplete, as there is no definitive account of what is out there in the mechanical world. It would also be immediately obsolete, as the process of innovation, obsolescence, manufacture, and discarding is constantly in motion. One proxy that gives some sense of how different things are, and how they are changing at present is in energy consumption per capita. This is not evenly distributed across the world. Highest per capita energy production is concentrated in those countries bordering the NAPT.

6.6.3 Knowledge and Memory

Knowledge and memory have increased exponentially during the same Late Holocene timeframe. The primary characteristic of Industrial knowledge is its structure. Humans are no smarter or more creative now than they have been in the past (leaning aside the conversation around the "cognitive revolution" that may or may not have produced intellectually modern humans). However, there are several key transformations in the organization of intellectual products that have changed our collective relationship to knowledge. First among these is the invention of science: the generally agreed-upon, consistently structured, and fairly reliable business of assembling and disseminating knowledge about the physical world. Other types of knowledge preceded science, and the collective success of human enterprises – and failures large and small – over the past 195,000 years bear witness to this. The unique characteristic of the modern scientific enterprise is the standardization of knowledge. That is, it is based on a standard language – down to the granularity of standardized letters, numbers, spellings, mathematical symbols, notations and so on. The extent to which standardization prevails in human intellectual and creative endeavors is easy to overlook. Yet everything from the language we use, which is standardized in dictionaries, to the 12 musical notes we play in the Western tradition, is built on the principle of universal interoperability. This principle of interoperability is relatively new.

The first English language dictionary was Robert Cawdrey's "A Table Alphabeticall", published in 1604 (British Library n.d.). The first edition of the definitive Oxford English Dictionary was completed between 1844 and 1928 (Oxford English Dictionary n.d.). After a preliminary edition in 1687 the first definitive French dictionary, "Dictionnaire de l'Académie françoise dedié au Roy" was published in 1694 (Calssiques Garnier n.d.). The metric system was formally created on June 22nd 1799 with the deposition of two platinum standards for the meter and the kilogram respectively in Archives de la République in Paris. In 1874 the Convention du Mètre was signed, founding the Bureau International des Poids et Mesures (BIPM n.d.).

The extent to which standardization permeates other areas of life also. A = 440 hz was generally accepted for western-style orchestras after about 1950. Prior to that tuning was much more varied, though there was a general trend towards standardization from around the nineteenth century onwards (Haynes 2002). The extent to which this standardization trend permeates late Holocene cultures is extraordinary. Thus, after most of human history has been spent exploiting a diverse array of scavenged and hunted foods, the agricultural revolution of the Greenlandian and Northgrippian boundary reduced farmer's food choices to predominantly domesticated and cultivated species. Within individual cultures there was a trend towards genetic standardization, measurement of portions, and control of consumption. However, the late Meghalayan is characterized by an acceleration towards standardization with no historical precedent. Bavaria's Reinheitsgebot beer purity law (1516), one of a number of medieval trading standards, is an example of the desire to establish some sort of quality control in the public marketplace. However, the influence of such standards pales beside the increasing standardization and order emerging from the regulation, information, and analysis associated with industrial food production. From the nutritional details on the back of food packaging, to the uniformity of process imposed by cook-books, largely unknown until the nineteenth century, moves us towards a commonality of experience. All of these standards, and many other have been accepted in varying but generally increasing degrees. Consistently one of the trends of the late eighteenth, the nineteenth, and the twentieth centuries was a growing trend of standardization and interoperability in knowledge systems.

Closely connected to the standardization of knowledge was the formalization of knowledge retrieval, that is, memory. Library catalogs are as ancient as libraries, but typically in the pre-modern era they were specific to each library. In the mid nineteenth century, inspired by the writings of Francis Bacon, Melvil Dewey, drew on the pioneering work of Natale Battezzati to develop his pioneering Dewey Decimal system, first published in 1876 (Wiegand 1988). The general principles of knowledge organization and retrieval established in the Dewey system provide an essential theoretical foundation for contemporary electronic systems. Universal cross-tabulation and rapid retrieval are the hallmarks of a successful library catalog, and indeed of any other data retrieval system. I has been possible for a literate person with access to a decent library to navigate and retrieve the core knowledge sets of human society for most of the past 150 years or so.

6.6.4 Energy

Possibly the most dramatic change in culture has been the transformation of human energy use. As in all other areas of cultural transformation this can be seen across the entire spectrum, from the societal to the deeply personal. Thus, our entire cultural pattern of energy use is currently as historically extraordinary as is our personal pattern of energy use. Whether we consider the millions of kilocalories of energy pointlessly consumed collectively in the simple business of heat output during electrical transmission, or the hundreds of calories of energy pointlessly consumed individually with each can of soda we drink, the conclusion is much the same: humans have more energy in controlled circulation at the end of the Holocene, both per capita, and in total than at any time in previous history.

As with other significant changes it is not only the amount of energy used, but also the types of energy used, and the ways in which it is expended that are historically unprecedented. As in previous cases it is not practical to make a complete inventory of all areas of change. Rather we must look for effective and accurate proxies. One area where the daily impact of changed energy patterns is immediately apparent is in the banishing of the night. Effective and near-universal night lighting is only about four generations old. Although humans have used fires, candles, torches, and lamps for much of human history nights as brightly lit as days are a complete novelty. So much so that they are an effective and novel evolutionary driver not only of human behavior, but also of other species. Insects, birds, and plants are all now driven by the eternal daylight of city life. Other examples of energy changes can be seen in the Urban Heat Island effect – the two to three degrees centigrade warmer cities are that the surrounding countryside, as a result of the combined energetic effects of active urban heat output in the form of mechanical activity and the concertation of animal warmth present in crowded spaces. Also, there are a number of passive effects, resulting from such things as the waterproofing of land surfaces, so eliminating the evaporative cooling effect common in non-urban areas.

6.7 The End of the Holocene

Humans evolved, survived, and prospered in the Earth's current and ongoing icehouse phase. Despite some fairly significant fluctuations in the extent of the ice, it has always been a part of our world. Although we were initially tropically adapted, all human evolution has occurred in the context of these icehouse conditions. While a lot of the discussion in this book focuses on the impact of both local and global changes on humans and their environment it is important to remember that some key components of the system have functioned consistently throughout the Quaternary. The basic structure and fabric of ecosystems has been consistent throughout. That is, there have always been tropical rainforests, deserts, temperate forest zones, tundra and arctic landscapes. There have been tropical oceans, abyssal plains, estuarine environments, and deep trenches. The locations and total area of some of the more malleable components of these systems have changed, but the core ecological elements have always survived. The broad patterns of ocean and atmospheric circulation have been the same. The distribution of land and water has been essentially the same.

Within most terrestrial ecosystems humans have encountered a fairly consistent cast of characters. In Africa and subsequently Southern Asia these players coevolved with humans. In other places where humans were newer arrivals they were direct competition. Terrestrial mammalian and reptilian apex predators were frequently drawn from the same handful of families, and where they were not, they were close analogs. These families include big cats, dogs, bears, and constrictor snakes. Similarly, terrestrial megafaunal herbivores have been remarkably consistent through this period, and often as hazardous to human health. For example elephants, rhinoceroses, bovines, and horses are all represented throughout the period. Likewise, members of the superorder Xenarthra, which currently includes anteaters, sloths, and armadillos, and historically included giant ground sloths and glyptodonts, have all coevolved, competed, or succumbed to humans throughout their shared histories. There are notable exceptions such as Australia, where these keystone families were absent when humans arrived around 40,000 BCE. However, dogs, cats, bears, and so on were substituted by an exotic collection of now mostly extinct marsupials occupying the same ecological niches. Prehistoric Australian megafauna included predators such as the marsupial lion Thylacoleo, the carnivorous kangaroo Hypsiprymnodontidae, the constrictor snake Wonambi, and some large marsupial plant-eaters such as the hippo-like Diprotodon.

Both freshwater and marine ecosystems were similarly consistent throughout the Quaternary, with the same families and orders of apex predators such as crocodiles, alligators, whales, seals, walruses, and sharks. Not all of these families are present on all continents, but if a human has encountered a rival predator, a megafaunal herbivore, or a potential prey species, it would be drawn from one of these families. Only in instances where humans have navigated to geographically isolated places, such as remote islands, have they found environments where these broad ecological patterns do not occur.

The consistency, while not seamless, points to an overarching structural similarity in ecological conditions throughout the Quaternary. Background extinctions, incremental evolutionary changes, and the catastrophic megafaunal extinction events outside Africa at the start of the Holocene notwithstanding, a broad range of trophic relationships have remained the same. There are three conclusions we can draw from this:

1. The biogeography of the Quaternary world has remained essentially stable through time, repeatedly returning to an equilibrium state on both land and water.
2. Despite periodic climate fluctuations sufficient refuges have remained for the major ecological building blocks to survive.
3. There have been no externally driven catastrophic events of sufficient magnitude to completely "upset the apple cart"

To put it another way, on a macro scale of time and space the genus Homo has enjoyed a period of ecological stability for its whole time on Earth

There is little doubt that we are currently witnessing a boundary event. The big question is, where in the chronostratigraphic hierarchy will this boundary ultimately be positioned? Is this the end of a stage, an age, an epoch, or a period? This seems to still be open for debate. Some boundaries unfold over decades, others over centuries or millennia. The refilling of evolutionary spaces once there has been a catastrophic clear-out can take much longer than the boundary event. The final status of the change in the geological column is often only finally established one the new chapter of Earth history is firmly established. In this instance we are in a historically unique moment, when we are aware of the mechanisms at work, we exercise some agency over some of them, and we can make reasonable predictions about how some of the processes may play out. My final thoughts on this I have relegated to the Epilogue.

References

British Antarctic Survey (n.d.) https://www.bas.ac.uk/media-post/deep-channels-link-ocean-to-antarctic-glacier/. Accessed 28 Feb 2022

British Library (n.d.) https://www.bl.uk/collection-items/robert-cawdreys-a-table-alphabeticall. Accessed 28 Feb 2022

Bureau International des Poids et Mesures (n.d.) https://www.bipm.org/en/about-us/. Accessed 28 Feb 2022

Camuffo D, Bertolin C (2012) The earliest temperature observations in the world: the Medici network (1654–1670). Clim Chang 111(2):335–363

Classiques Garnier (n.d.) https://classiques-garnier.com/dictionaries-of-the-academie-francaise-17th-to-20th-century.html. Accessed 28 Feb 2022

Deutscher Wetterdienst (n.d.) https://www.dwd.de/EN/research/observing_atmosphere/composition_atmosphere/hohenpeissenberg/start_mohp_node.html. Accessed 28 Feb 2022

Grill G Lehner B Thieme M Geenen B Tickner D Antonelli F Babu S Borrelli P Cheng L Crochetiere H Macedo H E (2019) Mapping the world's free-flowing rivers. Nature 569(7755):215–221

Haynes B (2002) A history of performing pitch: the story of 'A'. Scarecrow Press, Oxford

Karlén W, Fastook JL, Holmgren K, Malmström M, Matthews JA, Odada E, Risberg J, Rosqvist G, Sandgren P, Shemesh A, Westerberg L (1999) Glacier fluctuations on Mount Kenya since ~6000 Cal. Years BP: implications for Holocene climate change in Africa. Ambio 28(5):409–418

Milgram S (1967) The small world problem. Psychol Today 2(1):60–67

Oxford English Dictionary History of the Oxford English Dictionary (n.d.) https://public.oed.com/history/. Accessed 17 May 2022

Purvis A, Orme CDL (2005) Evolutionary trends in body size. In: Carel JC, Kelly PA (eds) Deciphering growth 2005. Springer, Berlin, pp 1–18

Repschläger J, Weinelt M, Kinkel H, Andersen N, Garbe-Schönberg D, Schwab C (2015) Response of the subtropical North Atlantic surface hydrography on deglacial and Holocene AMOC changes. Paleoceanography 30(5):456–476

Royal Botanic Gardens, Kew (n.d.) https://www.kew.org/science/state-of-the-worlds-plants-and-fungi. Accessed 28 Feb 2022

Udisti R, Becagli S, Castellano E, Mulvaney R, Schwander J, Torcini S, Wolff E (2000) Holocene electrical and chemical measurements from the EPICA–dome C ice core. Ann Glaciol 30:20–26

UNESCO Intergovernmental Hydrology Program (n.d.) https://www.un-igrac.org/news/global-groundwater-depletion-leads-sea-level-rise#:~:text=The%20hydrologists%20estimate%20that%20from,to%20283%20km3%20per%20year. Accessed 28 Feb 2022

US Bureau of Transportation (n.d.) https://www.transtats.bts.gov/TRAFFIC/. Accessed 28 Feb 2022

US Department of the Interior (n.d.) https://www.rivers.gov/map.php. Accessed 28 Feb 2022

USGS (n.d.) https://www.usgs.gov/news/usgs-high-plains-aquifer-groundwater-levels-continue-decline. Accessed 28 Feb 2022

Wiegand WA (1988) The "Amherst method": the origins of the Dewey decimal classification scheme. Libr Cult:175–194

Epilogue

Thoughts and Hopes

It was my original intent to end this book with the Late Holocene Boundary Event. Having documented the Meghalayan bounding stratotype there didn't seem to be much left to say. However, after over 2 years of writing about this, and try to explain what I'm working on to anyone patient or indulgent enough to sit around and listen, I feel obliged to respond to the question that almost invariably crops up in such conversations: "What can we do about this"?

My first answer is that if I could answer that question I would not be writing this book, I would be fixing the problem. Beyond that what I can share are the sources of hope and optimism that I have found. First and foremost, having reached the conclusion that we are in fact living through a boundary event as an Earth scientist I am profoundly excited. This is the first time anyone has ever documented this process in real time. It is an extraordinary opportunity to learn how the complex interlinked systems that define the Earth's surface functions under catastrophic duress. The instrumental record we make and the models we develop of how the end of the Holocene plays out will be unprecedented, and fascinating. Assuming there's someone around to look at the results of the work.

There are sources of inspiration, of course. It is tempting to fill a few pages with the inspirational words of thinkers and activists young and old who understand what is going on, and are working with diligence and intelligence to manage the changes, to convince people, communities, and decision-makers that action is needed immediately. However, others have taken this work on already with greater knowledge and expertise than I possess. Writers such as Heyhoe (2021), Goodall et al. (2021), Attenborough (2020), McKibben (2019), Newitz (2021), and Mann (2021) have addressed the core questions of policy, organization, and action with more eloquence than I can bring to bear om the topic.

M. Bampton, *The North Atlantic Polar Triangle*, Springer Polar Sciences, https://doi.org/10.1007/978-3-031-27264-6

My own rather bleak perspective is that the juggernaut of modernity, and most particularly those who control the greatest accumulations of wealth, and hold the positions of greatest power and influence, have shown little inclination to respond in any meaningful way to the impending changes. The Climate Change Conference (COP27) has, at the time of writing, just wrapped up in Sharm el-Sheikh, Egypt. No great policy statements have yet emerged from this meeting of the "Great and the Good". Indeed, any proposed changes would have to be so monumental, so totally transformational, that it is hard to imagine what that would look like, or how they would work. A solution demands at the very least a reduction of the world's population by orders of magnitude within a couple of generations; a total restructuring of our energy environment, and; an unpicking of the vast and complex network of manufacturing, consumption, and disposal that comprises much of our daily use patterns. Such action would also demand the abandonment of huge areas of land, and the return of large portions of the Earth to an unmanaged state. It is easy to say these things. But can I imagine living long enough to help deliberately build such a world? In a word: no.

If, as I have argued, we are living through a boundary event, all of these things will happen. But they will occur in the midst of a catastrophic change that triggers a societal collapse. The survivors will not be billionaire bunker-dwellers. For all of their resources they have failed to do the biological arithmetic to figure out how they will remain genetically viable in concrete silos for more than a couple of generations. Nor will they be those who have hidden away carefully-prepared stocks of tools, weapons, and supplies. For all of their ingenuity their preparations are based on a lifeline to a technologically sophisticated complex civilization, where such things as a reliable power, a well-stocked spare-parts bin, and a supply of antibiotics that are still in good condition can be counted on.

The survivors will be, as they always have been, those lucky souls who happen to be in the right place at the right time, and who have some combination of useful skills and attributes that enable them to manage the as-yet unknown combination of environmental challenges they will face.

The discussion of change at this scale has been a perennial component of popular consciousness, from the cartoon trope "Prophet of Doom" and his sign "The End is Nigh!" to the Biblical Book of Revelation. In some cases, there are explicit recommendations – from how to can produce to the appropriate spiritual preparations to "meet thy maker". In other cases, there are detailed analyses of the whys and wherefores of the process. There is now a recognized area of academic discourse frequently called "collapsology" (Diamond 2021). A recent and lucid exploration of this field is presented by Servigne and Stevens (2020), who note "we lack an overview of what a collapse might look like, how it might be triggered, and what it would imply in psychological, sociological and political terms *for the present generations*. We lack any real applied, transdisciplinary science of collapse" (Post Carbon Institute n.d.). I am not trained as a policymaker. I do have some limited material means to make significant changes to the way human society interacts with the environment. But no matter how frequently I cycle to work, how diligently I reuse and recycle, how strategically I vote, and how detailed my research, I will not change the historical trajectory of humanity, or the dynamics of environmental change. As I have argued in this book most of the trends can now clearly identify in the data were centuries, or even millennia in the making. Environmental history teaches us that the boundaries between chronostratigraphic units are only very rarely instant. Much of the time the changes take hundreds, thousands, or even millions of years. The bigger the boundary, the larger the required change, either in time or in magnitude. Thus, the mutually assured destruction threatened during the Cold War (and still a possibility in a nuclear-armed world, particularly as conflicts are either underway, or are threatened from Ukraine, to the South China Sea) would have created a clear mark in the geological record, and probably have defined a new Age. What we currently face is more probably a new Epoch.

Within the model of geological time the survival or perishing of humans is not really a concern. Our presence as agents of transformation is no more or less significant that the role of cyanobacteria or terrestrial vascular plants, both of which successfully transformed the composition of the atmosphere to a sufficient extent to ensure the definition of a new geological period.

It is possible to imagine a course of human action that might enable us to defy the evolutionary odds, and survive beyond the average mammalian species "life expectancy" of around one million years. We are capable, with a little care and a little luck, of joining the very exclusive club of species which have endured for tens of millions of years. However, accomplishing this would require some quite dramatic reimaging of how we interact with the world around us, and of who and what we really are.

References

Attenborough DA (2020) Life on our planet: my witness statement and a vision for the future. Grand Central Publishing, New York

Diamond J (2021) Collapse: how societies choose to fail or succeed. Penguin, London

Goodall J, Abrams D, Hudson G (2021) The book of Hope: a survival guide for trying times. Celedon, London

Heyhoe K (2021) Saving us: a climate Scientist's case for Hope and healing in a divided world. Simon and Schuster, New York

Mann M (2021) The new climate war: the fight to take Back our planet. Blackstone, London

McKIbben B (2019) Falter: has the human game begun to play itself out? Henry Holt, New York

Newitz A (2021) Four lost cities: a secret history of the urban age. Norton, London

Post Carbon Institute (n.d.) https://www.resilience.org/stories/2020-08-10/how-everything-can-collapse-excerpt/. Accessed 28 Feb 2022

Servigne P, Stevens R (2020) How everything can collapse: a manual for our times. Wiley, London

Index

M. Bampton, *The North Atlantic Polar Triangle*, Springer Polar Sciences,
https://doi.org/10.1007/978-3-031-27264-6

CPSIA information can be obtained
at www.ICGtesting.com
Printed in the USA
LVHW051943060723
751686LV00005B/54

9 783031 272639